AGAINST THE CURRENT
Organizational Restructuring of State Electricity Boards

Centre de Sciences Humaines (Centre for Social Sciences and Humanities): Created in New Delhi in 1989 the CSH, is part of network of research centres of the French Ministry of Foreign Affairs. The Centre's research work is primarily oriented towards the study of issues concerning the contemporary dynamics of development in India and South Asia. The activities of the Centre are focused on four main themes, namely: Economic growth and sustainable development, International and regional relations, Institutional structures and political constructions of identity and Urban dynamics.

(Centre de Sciences Humaines, 2, Aurangzeb Road, New Delhi 110011, India, Tel: (91 11) 301 62 59/301 41 73, Fax: (91 11) 301 84 80, E-mail: public@csh-delhi.com, Website: http://www.csh-delhi.com)

Institut Français de Pondichéry (French Institute of Pondicherry): Created in 1955, the IFP is a multidisciplinary research and advanced educational institute. Major research works are focusing on Sanskrit and Tamil languages and literatures—in close collaboration with the *Ecole Française d'Extrême-Orient*—ecosystems, biodiversity and sustainability, dynamics of population and socio-economic development.

(Institut Français de Pondichéry, 11, Saint Louis Street, P.B. 33 Pondicherry 605 001, Tel: (91 413) 334 170/334 168, Telex: 469224 FRAN-In, Fax: (91 413) 339 534, E-mail: ifpdir@ifpindia.org, Website: http://www.ifpindia.org.)

AGAINST THE CURRENT

Organizational Restructuring of State Electricity Boards

Edited by

JOËL RUET

MANOHAR

CENTRE DE SCIENCES HUMAINES

2003

First published 2003

ISBN 81-7304-478-3

Published by

Ajay Kumar Jain for
Manohar Publishers & Distributors
4753/23 Ansari Road, Daryaganj
New Delhi 110 002

Typeset at

Digigrafics
New Delhi 110 049

Printed at

Lordson Publishers Pvt. Ltd.
Delhi 110 007

Contents

PART III: LEARNING ON 'MODELS'

Preface

In order to improve the health of State Electricity Boards (SEBs), measures like corporatization, unbundling, and setting up of regulatory commissions have been undertaken by the Government of India. However, these options have proved insufficient in the states which have experienced them. Therefore, privatization has been encouraged but that too has achieved limited success, since:

— both the above reforms and the bidding phase in the privatization process take time, and
— the investors do not want to commit financial resources due to lack of information and guarantees on the means and latitude they will have to internally restructure.

Scholars involved in reforms, from the IIMs Ahmedabad and Bangalore, IIT Kanpur, ASCI, Hyderabad, IPE Hyderabad, and CSH, New Delhi, have evaluated the current initiatives and outlined some action plans to be further studied or implemented, in order to overcome the present impasse. These are presented and discussed in this book; issues in the internal organization of SEBs, as well as in their relationships with external bodies have both been addressed.

Joël Ruet

Contributors

PUNEET CHITKARA is Lecturer at the Indian Institute of Technology, Kanpur. He has worked with the Central Electricity Regulatory Commission in New Delhi, and his research interests are Energy Economics, Regulatory Economics, Applied Econometrics. He has contributed to the Indian 'Infrastructure Report 2002'.

GAJENDRA HALDEA (IAS) is Chief Advisor and Head of the Centre for Infrastructure and Regulation, National Council for Applied Economic Research, New Delhi. He has been Joint Secretary, Department of Economic Affairs, Government of India, and he has recently participated in the Expert Group on Settlement of SEBs' dues for the Ministry of Power, Government of India.

PREM K. KALRA is Professor, Department of Electrical Engineering, at the Indian Institute of Technology, Kanpur. His areas of interest are Power Systems, Expert Systems Applications, HVDC Transmission, Fuzzy Logic and Neural Networks Applications. He has contributed to the 'Indian Infrastructure Report 2002'.

R.K. MISHRA is Professor and Dean at the Institute of Public Enterprise, Hyderabad. He teaches and researches on Finance, Restructuring and Corporate Strategy. He has extensively worked on the Power Sector.

SEBASTIAN MORRIS is Professor at the Indian Institute of Management, Ahmedabad. His interests include international trade and investment, public enterprise, development economics. He has edited the 'Indian Infrastructure Report 2002'.

USHA RAMACHANDRA is on the Faculty of the Administrative Staff College of India, Hyderabad. She works on Power Sector Reforms,

Restructuring and Regulation of Infrastructure, Setting Power Tariffs, Regulatory Reforms.

D.V. RAMANA is Professor of Finance at the Xavier Institute of Management, Bhubaneswar, and consultant for Price Waterhouse.

V. RANGANATHAN is Professor at the Indian Institute of Management, Bangalore. He has been member of the Central Advisory Committee at the Central Electricity Regulatory Commission. He has been associated with reforms in Nepal and India.

JOËL RUET is Head, Department of Economics, at the Centre de Sciences Humaines, New Delhi, and Researcher, CERNA, Ecole des Mines, Paris. He works on the reform of the Indian public sector, in the fields of energy, water and sanitation.

RAJIV SHEKHAR is Professor at the Indian Institute of Technology, Kanpur. He has contributed to and is the co-editor of the 'Indian Infrastructure Report 2002'.

SIDHARTH SINHA is Professor, Indian Institute of Management, Ahmedabad. He has taught at University of Massachusetts, Amherst. His works are in the areas of takeovers, share repurchases and capital structure. He has contributed to the 'Indian Infrastructure Report 2002'.

1

Overcoming the Power Trap

JOËL RUET

State Electricity Boards (SEBs) are at the heart of the Indian electricity system, which should be in a position to supply electricity to a billion people. Sadly, there has been a rapid decline in their physical, organizational, financial, and moral condition. They supply electricity in a shoddy and ever-worsening manner. The main reason for this is that no amount of increase in the absolute level of generation can keep up with the steadily expanding social and economic demand. Whatever the ideological position on that issue, what ultimately matters is the simple fact that SEBs no longer provide electricity to its people, be they political clients, industrial users, or ordinary customers. It seems that the situation has, in fact, deteriorated after the process of reforms which have now completed a full decade.

Where are we today vis-à-vis reforms? And, more importantly, how does one proceed further? The main premise of this book is that, contrary to appearances, India has learnt considerably all these years—both in the economy as well as in the political economy of the power sector.

The power sector has two dimensions: political and technical. As far as economics must necessarily remain based on a 'technical' corpus, albeit politically discussed, it is the sincere belief of this book that, certain people, be they called experts, analysts, or economists, have an important role to play in the disentangling of various issues at the heart of the power sector.

This book gathers people from a large spectrum and solicits their ideas as to what would constitute tools for effective reform. Ideologies, for the most part, remain in the background. The reader can rest assured that the conclusions placed here rest, as much as possible, on articulated facts, experiences, and demonstrations; hypotheses, whenever they have to me made, will clearly appear as such.

Currently, we are still part of a learning process. The aim of this book is to make a contribution to a lively and naturally controversial debate; hopefully, certain proposals will gain support in the near future. I also hope that in a decade, if some of the contributors are right, the central problematic of the book would have substantially evolved from its present shape.

1.1 BEYOND CONTEXTUALITY: STRUCTURES MATTER

In order to improve the health of SEBs, the Government of India is encouraging corporatization, unbundling, and setting up of regulatory commissions. However, these measures have proved insufficient in the states which have experienced them. Consequently, privatization has been promoted. But privatization itself has turned to be deceptive due to certain factors:

— both reforms and the bidding phase in the privatization process take time,
— the investors do not want to commit financial resources due to lack of information and guarantees on both the means, and the latitude, that they will have to internally restructure.

In this context, it is argued here that the analyses available to reformers have reached their limits, and that a more systemized organizational approach should be formulated. Indeed, there is a crisis of production each time when the gap between demand and supply increases. There is a financial crisis when the SEBs generalize their arrears of payments to large branches of the public sphere. There is a confidence crisis when the Union Government and a few private investors demand more and more guarantees. When industrialists turn to captive generation, there is a crisis of maintenance and internal functioning; power cuts increase to the point of becoming endemic.

It is only very recently that the latter is being reported and analysed in systemic or structural terms, and is no longer dismissed with conceptually vague phrases like 'political day-to-day interference'. It begins to appear that the power sector is also structured by its day-to-day decisions in operation, maintenance, development and 'commercial', or rather less commercial, practices. And that these are taken and executed by people embedded in organizations which allow them certain degrees of liberty, and provide them certain managerial tools. Rigorous tools exist to study this phenomenon; the huge corpus developed by Herbert Simon or James March, and their followers immediately springs to mind. Further, an understanding of how organizations constrain choices within SEB, allows us to understand how they behave outside, with the public, with the state, with the Regulatory Commissions, with the much-in-demand private investors, who have so little knowledge about and confidence in the entity called the SEB.

On a pragmatic plane, these issues have a tremendous impact. It is enough here to refer to the withdrawal of one foreign private company in Orissa, and the acknowledgement of wrong expectations by an Indian company in the same state. Not to mention the failures of other privatizations, and reforms in the public sphere. Knowing what SEBs are and do, taking stock of previous experience, of good and bad surprises, helps in seeing the future.

Although these arguments are for specialists and practitioners, their implications for public debate should not be underestimated. The present situation generally receives two kinds of explanation. The book challenges both of them. The first among them is that the (public) electricity sector in general, and the SEBs in particular, mostly lack investment capacities. The other reason put forward is the generalization of 'political interference', that is, the interventions of the states not only in tariffs, but also in investment, or in day to day decisions (clientelist hiring and promotion, political protection in case of non-payment of bills, etc.). As soon as we believe in these sole simplistic perspectives, the solution appears clear enough. One should open the development of this sector to the private sphere, if need be foreign, and remove the Government completely from its management. This can be done through structural reforms such as corporatization, de-integration and the setting up of Regulatory Commissions.

What has actually happened?

The opening up of the sector has actually been gradually implemented since 1991. But a very pertinent problem—for the supporters of the financial liberalization hypothesis—is that this opening up was a failure. Few investors turned up for generation—the only segment opened up initially, with the SEBs remaining in charge of distribution. On the other hand this reinforced the supporters of a more ambitious and radical line—the general opening up of the power sector through the disintegration of the SEBs and the privatization of the distribution process. While this enabled the two lines of analysis to mutually strengthen each other, it also contributed to a certain blurring of issues—structural reforms were longer separated from privatization. But here too, the investors did not make a beeline and the few experiments undertaken or finished give reason to believe that they will not come or not as one would like them to. In the electrical sector, it is still 'waiting for Godot'. One is thus faced with a dominant analysis, which is undermined by the empirical failure of its predictions.

These current lines of analysis are presented and challenged in detail in the first part of the book and provide two views on the sector, the first by Ruet (Chapter 2) and the second by Haldea (Chapter 3).[1]

However, in this book, the authors underline what many field studies prove. That is considerable room for straightforward improvement is available, with limited financial means. Mishra (Chapter 4), Ranganathan (Chapter 5), Ruet (Chapter 6), Kalra et al. (Chapter 7), Morris (Chapter 8) and Ramana (Chapter 11) articulate several internal proposals for realizing this. As far as the external environment (financial regulation, links with the State, with the Regulatory Commission, scope for private involvement) is concerned, Ruet (Chapter 6), Morris (Chapter 8), Ramachandra (Chapter 9) and Sinha (Chapter 10) give details and propose both accompanying institutional mechanisms as well as general frameworks.

What comes out of all this, to be detailed in the next subsection, is that neither the argument of the absence of funds, nor the notion of 'interference' reveal the complexity of the procedures at work in a SEB. They only characterize SEBs 'in hollow', in the sense that they describe the non-adequation to a norm, and by doing so they rather describe the norm itself than what actually exists. They do not allow us to imagine what could happen if a SEB had money, or no politicians interfered in its workings. In that sense, they do not describe the reality, and do not allow dealing with predictions.

The essays in this book essentially constitute a search for articulated proposals to reverse the internal rationality of SEBs. And its belief is that accumulated experience has now grown enough and reached critical mass.

1.2 LOOKING EITHER IN NEW DIRECTIONS, OR WITH A RENEWED PERSPECTIVE

1.2.1 TOOLS AND MEASURES

Mishra, Ranganathan, Ruet, Chitkara, Kalra and Shekhar, and Morris propose reform measures that are organized in a gradation, from measures that can be applied within an environment remaining public, to a private environment design. The gradation in these measures can be regarded as a range of available tools, to be selectively picked up by governments according to their desired degree of reform. But it can also be considered as a kind of phase-out programme, starting from the easiest and going on to the deepest reforms, for they are based on the same grounds presented above.

Mishra analyses the present use and anticipated advantages of internal audit (IA) which comprises audit of operations concerning finance and accounting, administration and management, operations and services, and allied matters in nine power utilities in India. He provides the backdrop, scope, charter, coverage, reporting arrangements, and performance of the internal audit in these utilities. He shows that however well-identified a tool might be, its implementation varies as far as audit charters, committees, and IA plans are concerned; he also underlines some good practices. His contribution makes clear that, to be an integrated tool in decision-making, and as a feedback tool for the management, IA should include the functioning of the various controls, and not be limited to the coverage of operations alone. He points out the existence of recommendations originating in IA departments, that are accepted by the hierarchy, without being further implemented. The given explanation is the lack of funds, but again an organizational analysis could reveal that structural lock-ins exist. This proves that analytic competence exists within SEBs, at least more than is usually agreed.

Ranganathan focuses on the setting-up of profit centres under a renewed approach. The classical argument against these is that, structurally, because of the load mix (that is, the share of respective consumers with different tariffs), some zones are by definition not profitable. Drawing on the experience of Nepal and Maharashtra among others, Ranganathan manages to overcome this contradiction by proposing a scheme articulated on creation of information and focused on operational costs. He introduces the principle of cost centre and surplus centre, which allows, through progressive implementation any zone to become profitable by construction. He also rightly insists on organizational details that decide the ultimate success or failure of such tools. For instance, he deals with certain aspects of control systems like the cadre rivalry between the engineering branch and the accounts branch, and its impact on a key task like disconnection of consumers who have defaulted payment. These organizational details are known from studies but hardly discussed when it comes to implementation. In other words, Ranganathan proposes a phased-out implementing process that is sensitive to the internal organization.

Ruet, based on a study of seven states, gives a systematic description of the decision-making in SEBs by highlighting the structural organizational roots of inefficiency. He shows how, because of the administrative nature of the budgetary and information systems, decision-making cannot rely solely on the cost benefit analysis resident in them. He then advocates a phased-out scheme for allocation of budgetary power to the field officers, that is, the people who devise the day-to-day operational decisions. This goes beyond and encompasses the profit centre tool, and its implementation implies a learning process, which is taken into account in the various suggested control and monitoring systems. Keeping in mind the failures of the reforms within the public sector as well as the attempts at privatizing, an accompanying scheme based on a management contract is proposed. Again, taking into account the experience in this matter gained from Orissa, some specific tools and policies have been designed to cope with its inherent risks. The scheme plans to introduce financial flexibility in the operation of a 'private' incentivized human resource hierarchy based on voluntary participation.

Going one step further, and taking privatization as a reference point, Chitkara, Kalra and Shekhar propose an original and systemic approach for risk analysis and benchmarking techniques for SEBs. They underline the need for indicators regarding the inadequacy of both physical and human resource assets, and underline the revenue and regulatory risks. As far as the former is concerned, they provide a specific focus on metering, billing and collection, and for the latter, on assets valuation and actual payment of subsidies by governments. Combining this with power quality at the consumer's level, they argue for the need of a global indicator to measure the performance index. They also use their analytical framework to propose the private operation of auxiliary services, i.e. generation of 'reactive power' and hence the regulation of the voltage profile. So far these have been neglected in India though they are at the heart of a proper electricity quality. This measure could encourage confidence in privatization and achieve high results for low investment.

Writing from one extreme end of the gradation from public reforms to privatization, Morris argues for a quick privatization by largely focusing on a generation-distribution articulation in order to overcome most of the revenue and regulatory risks. Since, per se, this is not enough of a guarantee, he deals with all the implied restructuring and regulatory issues. He also advocates an incentivized human resource management, by stating that the courts should no longer consider distribution and generation companies as coming under the state's control. Taking the example of commercial measures, he argues that the necessary freedom is not available in government departments and public sector companies, which are more rigidly oriented towards rules. Thus, despite the fact that this set of measures is based and articulated on the case of Gujarat, it can be generalized and considered to apply to most of the Indian states.

The set of measures proposed by the book then offers an actual gradation in the depth of measures, but they are always given with an implementation agenda, and they can also be considered complementary to each other. Indeed, all of them shed light on the managerial, organizational issues, and show how these aspects alter, constrain, and veritably frame the decision-making towards inefficiency. The complementarity of such measures is shown and dealt with by Ruet in the concluding chapter of the book (Chapter 11). The diversity and modularity of these should offer scope for a specific adaptation to the practical as well as political conditions and aims of any given state.

1.2.2 Applying Tools: Three Case Studies

The book also provides in-depth documentation and analysis of the reforms till now in two states that can be considered, in many respects, as two of the most advanced in reform. These are Orissa, analysed by Sinha and, under a different perspective, by Ramana, and Andhra Pradesh, dealt with by Ramachandra.

Sinha offers a tariff and finance-based approach, but he also deals with the importance of examining privatization as a process.

He underlines the huge burden that still constitutes a regulation based on World Bank improvement estimates dating back from 1996, and not on present estimates of actuals. Echoing tools proposed in the previous part of the book, he recalls the financial burden of low collection in rural areas, and deals with the human resource aspects, especially vis-à-vis the monitoring of the management contract that was experimented with BSES in 1996. This echoes ideas expressed in chapters by Morris and Ruet. He points out the successful experiments of 'micro-privatization' at the village level, and emphasizes the need for a multi-year tariff programme, in which a phased-out transition period with remaining subsidies is unavoidable. Interestingly, the states of Maharashtra and Andhra Pradesh are now considering this; what began as a learning process has begun to materialize in new developments of reform.

The need for re-thinking reform is clear from the example of Andhra Pradesh, presented by Ramachandra. She discusses the process of transition, the organizational structure, the regulatory framework, and asks how to financially deal with the past functioning of the Board. This chapter highlights the major errors, not to be repeated again in any reform undertaken in any state. The most important among these have been the revision of the value of distribution assets (doubled!). This has been the solution adopted to cover the key pension fund liability only to leave it with the newly created companies. This in turn has lead to artificially high tariffs. Again, the provision that the personal might have a five years period to think out his status has contributed to absence of clarity in the reform. The present situation is that power purchasing costs are both unsustainable and unregulated, Andhra Pradesh can only go for the suggested multi-year tariff regulation and for regulating generation. But real internal reforms can only be kick-started if some of the tools offered here are actually implemented.

Finally, Ramana suggests a model based on the principles of cybernetics, the Viable System Model (VSM). He focuses mostly on distribution companies. He underlines that VSM offers a paradigm for problem-solving. Through the diagnosis of the limitations of the existing solutions he suggests the importance of seeing the organization as a Viable System consisting of several viable systems. Beyond the corporate level, such systems are conspicuous by absence at the division level. According to Ramana, each subdivision and division has to become viable.

What is striking is that while Orissa has pioneered reforms in India, often suffering on account of the trial and error process, Andhra Pradesh has all the time closely examined and gained experience from its neighbouring state. But in a way Orissa has also benefited; attracted by the state's reputation as an experimenting lab, private parties have invested heavily. States embarking for privatization these days no longer have this kind of appeal.

1.3 CONCLUSION

The public debate on power sectors reform has to change its orientation. The clichéd opposition between the public sector and privatization needs to be refined. This book not only provides certain tools, but it is also an attempt to anticipate and participate in the growing public understanding that the change in 'ownership' that actually matters is a change in the rights of control. That is, a change in the way decisions are taken—no matter who makes the investments—deemed to become profitable or at least sustainable. What matters is not that objectives are different from public to private, but that the way decisions are enframed in a structure is largely determining. The book provides directions to change this structure. In that respect, it disentangles usually mixed issues, technique and ideology, in a field where considerable technical change has led the sector to lose its status of a naturally integrated monopoly. It thus provides analytical clarity, which will hopefully lead to a sound political debate.

PART I

INTERNAL LOCK-INS, EXTERNAL FRICTIONS

2

The 'Enterprisation' of the State Electricity Boards

JOËL RUET

The poor financial and technical condition of the Indian State Electricity Boards (SEBs) is widely acknowledged. There is less consensus on its roots. Section 1 addresses this issue by disentangling the role of 'irrational tariffs' (as pinpointed by many blueprints for reform) from one of the SEB's own cost inefficiencies, i.e. 'irrational management'.

If one has to believe that the system is irrational, one question must however be raised: what can one do with it, i.e. with irrational people? Certainly nothing, as whatever the reform, irrational behaviour won't be efficient; people have to be changed, hence the credo for privatization. However, viewing people as irrational does not help to understand and solve the issue.

Section 2 adresses the rationality of a system that has sustained over fifty years. It focuses on the link between costs and tariffs, by analysing the decision-making process in SEBs. It is shown that SEBs have neither been asked to look for cost efficiency, nor given the means to do it. They have been administratively managed and not managed as firms.

Section 3 discusses the need for SEBs' 'enterprisation', (i.e. turning them into firms: the concept has been coined in the study of post-socialist countries, see Hirschhausen 1994; Bomsel 1995), which is a *sine qua non* condition to address their cost efficiency. The extent to which the process of present reforms is actually a process of enterprisation is analysed. It appears that enterprisation is, as of now, incomplete.

2.1 LOW TARIFFS BUT ALSO COST INEFFICIENCY

Tariffs do not cover the SEBs' costs, which leads among things to the inability to invest for cost efficiency and in new capacity. Tackling only this consequence without addressing its roots has failed as an approach. To explain this financial failure, SEBs argue against low tariffs. In contrast, state governments pointing out the cost inefficiency of SEBs, say that they do not want tariffs to be passed through to them. Since this diagnosis has an important bearing on the perception of the ability to reform and hence on the reforms to be proposed, let us examine why both arguments are correct.

2.1.1 The Financial Distress of SEBs

SEBs face recurring commercial losses. This financial situation has in turn led to the inability to pay the coal suppliers and the national generation companies. The sector has failed to maintain assets, postponed investments and has not developed new generation and network capacities to sustain the demand, both in the long and the short run. The energy shortage is estimated at 11 per cent. In 1991-2, to tackle the investment problem, the Government of India allowed foreign private participation in generation. Bulk tariffs ensuring a rate of return of 16 per cent were guaranteed to attract investment. But this rate did not appear credible to investors, as SEBs in poor financial health would be the buyers. Only 2.7 GW have been installed by private companies to balance the 20 GW gap in the 8th plan.

As a consequence of this failure, the Indian government as well as some state governments have taken stock of the major role played by low tariffs and low SEB efficiency. In 1996, the central government reached a formal agreement with various state chief ministers, stating that structural reform in tariffs and management is necessary, adressing the 'SEBs poor management' and the 'political interference in the day to day affairs of the board'.[1] Now, while SEBs argue for tariff increases, state governments prefer to focus on SEBs' inefficiencies in costs.

As far as tariffs are concerned, from the 1980s, subsidies through undercharging of the agriculture and domestic sectors have increased, primarily due to local political decisions. The ratio of average tariff to cost of supply is equal to 0.76. According to the Planning Commission, the effective subsidies to agricultural and domestic consumers are Rs. 15,300 and 3,900 crore respectively. Out of this only, Rs. 8,000 crore are balanced by cross subsidies from the industrial[2] and commercial sectors. Theoretically, the balance has to be paid by state governments through financial transfers to meet the statutory 3 per cent rate of return. This is, in practice, not always actually done. Indeed, only Rs. 2,600 crore are covered by financial transfers from state governments, with 10 governments out of 19 giving nothing. Thus the allegation that tariffs prevent SEBs from reaching financial equilibrium is technically not unfounded.

On the SEBs efficiency side, relative laxity in the metering of energy, and high levels of technical and 'non technical' losses, are noteworthy. The latter include thefts, bribing, unmetering. At times, these are instigated by SEB agents, at others they are tolerated by them. It is difficult to separate technical from non-technical losses in the absence of a proper metering system. The total level of losses remains partially unseen due to accounting manipulations. A part of the losses can be passed in the accounting books to unmetered users, who are charged flat rates. Also a part of the money collected from the high tariff consumers can be declared to come from low tariff consumers: this way the same amount billed corresponds to a higher quantity of energy billed and to a lower level of losses.

Nevertheless, on the basis of audits, World Bank-appointed consultants assessed the total level of losses at 40-50 per cent over India.[3] With this level, the final average cost is at least twice the generation cost. According to the interviewed people, a credible rough estimation of the technical losses seems to be about 20 per cent, i.e. almost twice the international standards. Therefore non-technical losses would amount to 20-30 per cent. This high level of energy loss contributes to the commercial losses of the SEBs. Its technical part shows that the investment share between generation and network has been far from optimal. Similarly, low operation performance (poor Plant Load Factor, high rate of burnt transformers) shows the low expenditure in maintenance.

So, the commercial losses of the SEBs can be explained by low tariffs as well as by high costs. Tariffs can therefore be interpreted as being far from their proper level; this is the SEBs' interpretation. But they can as well be seen as not so far from what would be proper costs under performing management, which is an argument used by state governments. The relations between these two aspects need to be analysed more precisely. The basic tariff per unit principle is a cost plus system: the SEBs have to recover their costs through the billed energy. Thus, the physically lost energy increases ex-post the per unit cost of supply. As SEBs hide a part of it through accounts, attributing it to flat tariffs or low tariffs users, it has to be repaid through financial transfers from state governments. This is the root of the conflict between state governments and SEBs.

To clinch the debate between these two 'interpretations', one needs to look into the black box called SEB, and elaborate on the actual status of the 'day to day interference' and its effects on technical and non-technical performances in SEBs. This paper shows how this 'interference' affects the costs and the level of tariff-setting.

2.2 SEBs ARE NOT FIRMS, i.e. THEY HAVE NEVER ASKED TO BE EFFICIENT, WHICH THEY MANAGED NOT TO BE

The low cost performance of SEBs is linked with political interference; they have not been so much 'poorly managed' according to a stated objective of economic efficiency, than managed according to objectives other than they were actually given. The SEBs' style of functioning is administrative and not the way firms are run. The point is not the juridical status of SEBs. While SEBs are constituted as administrations, they are commonly analysed in terms relative to firms. This section argues that this administrative nature is the very cause of SEBs' inefficiency, through a specific and very internally consistent system of incentives, that derives from the information system and budget constraints.

2.2.1 Decision Making in SEBs

The World Bank[4] argues that due to 'pervasive politicization of most decisions' SEBs were 'progressively transformed into an extension of the state government, unable to meet the power requirements of the state'. This leaves partially unanswered the inner logic of management. Indeed, the SEBs managers' objective has not been inefficiency *per se*. The state governments too have expectedly not agreed for deficits *per se*.

Investment affects technical efficiency and hence technical losses. Power sector is often viewed through its externalities on development. This is especially the case in rural electrification of developing countries. This has happened in India (cf. Chakravarty, 1987), from the mid-1950s onwards. Policy directions with corresponding budgets have been given to SEBs. In itself, this does not prevent the SEBs from behaving as firms, since the actual implementation and choice of techniques is left to them, based on cost of supply minimization. But the politically given objective was to electrify at the fastest rate, with given schedules. Lists of villages to be electrified were provided. This does not mean that prioritized zones within the state were selected, with the choice on the mode of electrification left to SEBs. Instead, in several zones, numerous smaller 'political stronghold' areas had been prioritized, keeping in mind electoral considerations. This severely constrained the choice for SEBs. They had no alternative but to electrify simultaneously widespread areas, using long low voltage (LV) lines. Being cheaper than medium voltage (MV), this effectively ensured quick electrification as more villages were reached in a short term by this way. Nevertheless, this low MV/LV ratio has generated a high level of technical losses and therefore financial costs to the SEBs. This in turn has ultimately led to higher cost of supply and/or less financial resources for electrification itself.[5]

Electricity networks have externalities on development. So it could be argued that the cost attached to this fast rate of electrification has been balanced by fast development of agriculture and helped the green revolution. Whatever the macro efficiency, the point here is that such a comparison, if it has been done, has been based on an objective different from the strict minimization of the cost of supply (years later it still affects the SEBs efficiency), and imposed on SEBs an external discretionary power. It is an administrative decision. Such an analysis is not antithetical to the vision of the power sector as a way to redistribute. Indeed, electrification, financed on budget, has redistributive effects. But, from the SEBs point of view, this policy direction doesn't affect its financial running. On the contrary, executive instructions have led to a situation where a part of this redistributive policy has been passed to them.

More generally, executive instructions are given on every investment, 'the criterium being not the additional revenue but the satisfaction of local needs', an official ventured to say. For instance, politicians put pressure for the location of new plants in their stronghold, ask for priority supply even if the collection is poor, or to allocate to it more credits on maintenance. Also, for electoral reasons, investments have been dispersed in many projects simultaneously instead of on prioritization, leading to costly delays. Just before elections politicians ask for visible immediate actions. Similarly, the low ratio on investments in network versus generation can be explained like this: first, generation is more visible as 'MW speak more than km of lines' both to populations and administrative controllers, and second, generation development implies more contracts than network and opens the door to more bribing.

Interference appears in recruitment, in promotion and also in agents' transfers. The SEBs had to recruit numerous unskilled workers, which increases the costs. Unskilled workers were recruited in politicians' strongholds regardless of the needs of SEBs, to such an extent that Rajiv Gandhi referred to them as 'State Employment Boards'. All hierarchical nominations (chairmen, board of directors, the intermediate hierarchy) are political. Potentially each nomination of a position in the SEBs can face political pressure to support one candidate or another.

Non-technical losses come from the commercial activity of distribution. It involves metering, billing and collecting. All three are inefficient in SEBs. But absence of meters, flat tariffs for agriculture, manual or external billing and poor collecting indicators do not explain everything. Thefts allowed by corrupt SEB agents and political protection for clientele, are also part of the explanation.[6]

2.2.2 SEBs are Managed as Administrations and not as Firms

SEBs are given political objectives and furthermore asked to fulfil several particular interests through direct executive instructions, i.e. an external discretionary power. They are therefore administrations ordered to fulfil political objectives. Indeed, according to the definition given by E.C. Blanfield, a firm has one main objective only. Its parameters are easy to measure and can be expressed in monetary terms. This can be profit for a classical private firm, or in the case of a public firm in natural monopoly, this objective may become the minimization of the long run cost of supply. It involves a highly integrated measurement and control system, with rules to solve objectives conflicts, based on economic criterion (evaluation of return, cost comparisons). Milgrom and Roberts add that 'a firm is a business organization with a central office that has substantial discretionary power' and has 'substantial independence from other discretionary powers as Courts, Governments, Agencies'.

Conversely, in administrations, according to Blanfield, there are several objectives of different kinds, that are at best unprioritized or may even be contradictory. So conflicts are not subject to resolution through a robust decision criterion. Decision-making ultimately depends on the use by the government of its discretionary power.

Accordingly, for an organization to be a firm, there are the two necessary conditions: (i) the existence of a main monetarily expressed objective—its derivative being the use of an adapted information system; and (ii) the absence of an external discretionary power. Should one of these conditions not be fulfilled, it would be enough to show that the organization is not a firm, whatever may be its official designation.[7]

The objectives outlined for the SEBs, are all alien to the concept of efficiency. The latter can therefore hardly be demanded of them; there are no incentives to pursue it.

2.2.3 A Consistent Self-Enforcing System

The information system is poor and does not allow hierarchical and external control on efficiency. At the board level, the total amount of losses—technical plus non-technical—is completely known, as the amount of energy generated or bought versus the amount of energy paid for is known. But the accountancy units are the divisions, which are supplied through substations issued from SEB transmission network. As there is generally no metering at this level, the actual consumption of each division and its actual efficiency remain unknown. In some places this information exists, but the absence of proper metering in each low voltage feeder, and at the users level, renders the proportion of technical and non-technical losses an unknown entity. The very basic management tool that enables internal audit or benchmarking is therefore not available. Each distribution unit can claim its innocence about losses or its relative innocence due to its backwardness.

Technical losses can only be estimated by SEB calculations, with no evidence that they reflect the real performance; from these are deduced the so called non-technical losses. In them, it is impossible to know the share of government policy, of politician's interferences and of the SEB's own inefficiencies. As state governments have their own financial problems, they complain about the SEBs' performance, but they do so on inadequate information. Hence in this paradoxical information system, officers cannot 'manage', while state governments are not able to fully control SEBs.

Let us now detail the tariff issue from the perspective of a soft budget constraint. SEBs do not control their tariffs and are only partially responsible for their costs, due to executive instructions. The profitability of a decision can therefore not be estimated ex-ante by economic analysis. As subsidies are subject to ex post negotiations and are not written in the state budget, SEBs are not sufficiently assured about retaining the ex post actual incomes of their decisions. Benefit is ultimately fixed through an administrative decision. SEBs then face a soft budget constraints, as defined by Kornai in his analysis of the socialist system. More precisely, they are called 'soft subsidies' since they are 'not expressly laid down for a longer period' and the 'amount (is) subject to bargaining'. Besides the incentives to follow political demands, SEBs managers have no hard financial constraints or incentives to integrate into their decisions.

Conversely, states' control on SEBs tends to be somehow lax. An official I interviewed explained that as states fix the tariffs and impose policies without actually paying subsidies, they have to 'extend loans and give help by way of capital subsidy'. He agreed that in spite of 'five or one-year plans, the budgets were not used for the purposes originally decided upon in consultation with the planning commission'. 'Theoretically, everything is under government control, but SEBs are in the red. The priority of the governments is the agriculture, so SEBs are not strictly monitored', he added. Another confirmed that in this situation operational efficiency was not an objective for SEBs, whose tendency was to ask the government for support through credits or simply to accumulate losses. A consultant agreed that 'as it was a priority sector government funds came anyway and were less of a worry for SEBs'.

Let us now systematically address the consistency of this administered system. Figure 1 details how (i) multiple objectives fulfilled through executive instructions, (ii) the use of criteria other than cost efficiency, (iii) soft budget constraints, and (iv) a poor information system, interact to sustain themselves.

It has been seen that executive instructions affect cost efficiency. These instructions are actually disincentives for SEBs managers as far as cost efficiency is concerned. Besides, economic analysis doesn't allow assessment of the profitability of investment. Indeed, the expected income of managers can change after any new executive

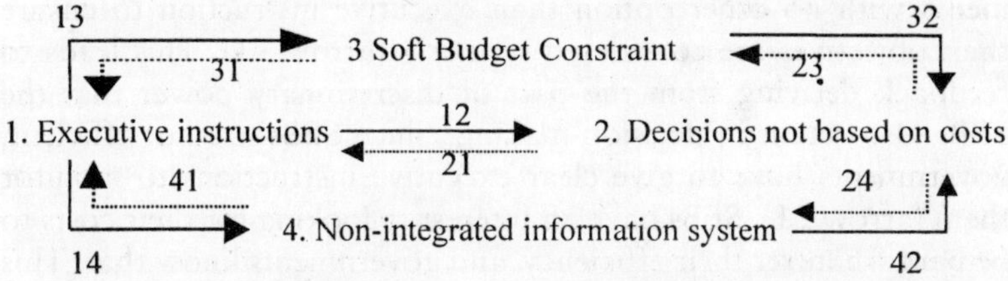

FIGURE 1: THE CONSISTENCY OF ADMINISTRATIVE CONTROL OF SEBs

instruction, not to mention the advantages that accrue to illegal consumers. For these reasons, executive instructions lead to the exclusion of economic efficiency from decision-making (arrow 12). At this stage of the essay, I would like to show how this direct causal relation is sustained by soft budget constraints and the information system.

Given the multiple objectives of governments, including redistribution, tariffs are the results of political trade-offs, affecting SEBs profitability ex post. This leads to soft budget constraint (arrow 13). Hence, there is a disincentive for SEBs in using a cost efficiency criterion, as their results are likely to be expropriated through the subsidy-setting process (arrow 32). Also, as politicians give executive instructions, they cannot strictly monitor SEBs; hence the formal control has to be weak. Indeed, the less the information, the less the costs of executive instructions is known, and the less they can be opposed (arrow 14). In this context, even if there were incentives in using a cost efficiency criterion, it would be hard to apply due to scarcity of information (arrow 42). This shows that behaviour of SEB managers is therefore consistent with the system; they are not *a priori* 'poor managers', in fact they are not a posteriori allowed to be better, given the incentives system. Their job consists in satisfying local needs, and the technical running of the electricity system. They could have done more but had no rational economic reason to that.

Conversely, there is some feedback (materialized in the figure through disjointed arrows) that strengthens the consistency of the system: as the cost is not the criterion, there is reduced need for an information system and for managerial tools that cannot be applied (arrow 24). Costs of decision being unknown and actual payment of subsidies being left for later, the politicians have space to give executive instructions with immediate political return (arrows 21 and 31). Poor information on costs sometimes leaves the governments with no other option than executive instruction to ensure their objectives are considered seriously (arrow 41). This leads to feedback deriving from the part of discretionary power that the SEBs also retain in decision-making. Since SEBs are not efficient, governments have to give clear executive instructions to monitor them (arrow 21). SEBs have an interest in looking for their costs to be paid whatever their efficiency, and governments know that. This is where the process described as soft administrative pricing comes in. This in turn sustains the soft budget process (arrow 23) leading governments to lower subsidies through executive instructions (arrow 31). SEBs also benefit from poor information which covers up areas where their inefficiencies are located (arrow 24).

As Kornai showed in the case of the socialist system, there is an 'affinity' among the elements of this managerial system; not only are they consistent with rational behaviour in it, but they also strengthen and legitimize each other.

2.3 REFORMS AND THE NEED TO 'ENTERPRISE' SEBs

As we have seen, looking for cost efficiency makes sense only within firms. Therefore, if there is a will to address the cost efficiency of SEBs, they have to be 'enterprised', i.e. turned from administrations into firms not only in their status, but also in their decision-making. Reforms need to enable SEBs to look at cost efficiency without any external discretionary power.

Enterprisation is a concept that has been coined in the study of post-socialist countries (Hirschhausen 1994, Bomsel 1995). It not only refers to an administrative starting point and a firm ending point, but it also refers to the processes that do that. This process is specific to the type of goods that are concerned (the type of the future firm market) and even more to the type of initial administrative system. In case of SEBs, this detailed process is still not defined. This will be a task for future normative work, based on the experience of the first reforming states. At this stage, the concept of enterprisation helps in examining whether, after structural reforms, SEBs have become firms. It can serve as a criterion for positive analysis of reforms.

Let us introduce why the current reforms are not expected to bring this (a detailed analysis based on case studies can be found in Ruet, 2001).

2.3.1 Corporatization

Corporatization is the modification of the juridical status of the SEBs, namely to move them from the status of boards, i.e. administrative bodies, to companies ruled under the Indian Companies Act of 1956. This involves accounting rules which clarify the financial results before subsidies and also enable us to know the exact level of subsidies really transferred. Also, corporatization agreements have to be signed between companies and governments to delineate their respective roles, rights and obligations. I will refer to this juridical change as 'juridical corporatization'. In reforms supported by the World Bank,[8] what is called corporatization includes more: it is expected to be done practically by training and consultancy on (a) system operation, (b) maintenance, (c) finance and accounting, (d) metering, billing, and collection improvement, (e) human resource, and (f) environmental management. Beside the juridical aspects, it is this complete approach that I will refer to as 'corporatization'. Corporatization is mainly intended to be effected through the implementation of modern techniques like new software and hardware; new accounting systems, including profit-centre and cost-centre accounting; a functioning information system, internal audit and so on.

Juridical corporatization would bring the SEBs under the Indian Companies Act 1956. This barely changes anything in itself for SEBs; they will have to issue annual reports on accounts, and directors will be appointed by the shareholders. Apart from calling the state governments 'shareholders', there is hardly any significant change. Corporatization as defined by the World Bank, on the other hand, brings something more: it mainly provides an information system, which allows benchmarking and can bring transparency. It makes available managerial techniques based on cost efficiency objective, which is a potential step towards enterprisation. However, there is no assurance that they will actually be used.

First, it will need time, as practices change slowly. For instance, whatever the ownership, the fight against corruption, the training of unskilled helpers and of SEB agents will be a time consuming process. Considering the present level of computerization, the capacity of information transmission in rural India, the infrastructure available, the resistance of unions and agents, the actual pace of entreprisation is hardly predictable. But what remains certain is that corporatization doesn't in itself ensure entreprisation. After juridical corporatization has taken place and after the consultants have left, the process must endure. There is asymmetric information between different hierarchical levels within the SEBs, between SEB insiders and outsiders, as also between different outsiders, about the potential rhythm of entreprisation.

Second, if the owners of corporatized SEBs continue to be state governments, they shall be tempted to maintain the soft budget constraint, and executive instructions might continue, even if their costs appear more clearly. They can also ask the corporatized SEB not to apply for a tariff revision from the commission. In other words, actual change partly remains linked with political commitment not to interfere.

2.3.2 Setting up of Regulatory Commissions

According to the Regulatory Commissions Act (1998), each state has the legal possibility to create a commission (SERC). Its role is to determine the accurate level of tariffs, and the setting of profitable (on a cost plus basis) operating conditions for the operators, provided they achieve actual improvements (loss reduction, quality improvement) for the consumers. However, it depends on each state government if it wants to attribute some other functions to the SERC. Beside tariffs, their role can range from a simple designer and controller of minimum technical standards and minimum information requirements, to a dispute adjudicator, or to a strong regulator, that is to approve investment and control licensees. Provisions in the Act on the selection and removal of SERC commissioners assure minimal autonomy vis-à-vis both central and state governments, industry and consumers. The commissions have all the powers necessary to design and enforce their orders. In the last analysis, the main delineation in the SERCs' credibility is the way they are set up: through a state bill or through a decree, as in the latter case they are easily removable. Indeed, tariff setting orders in Orissa (where the first such commission has been established) have proved to be transparent procedures.

Do regulatory commissions help to enterprise? Although they have no direct action on the way state governments influence SEBs, they provide strong constraints. SEBs have to issue affidavits for tariff settings. If they are found to be wrong after regulatory commissions enquiries, SEB managers can be prosecuted. This provides an incentive to keep information up to date. This has effectively happened in Orissa, Haryana and Andhra Pradesh, putting transparency on losses. Better information made available through corporatization is an important weapon for both SEB managers and state governments. As subsidies are booked, and procedures and annual targets given by regulatory commissions made more transparent, SEBs are in a better position to ask for subsidies and financial support from their shareholders. Invoking the pressure exerted by regulatory commissions, managers can argue with more confidence against day to day interferences. State governments can use the regulatory commissions to make SEBs face their responsibilities. So, there could exist some window for a public reform, i.e. enterprisation under public ownership, based on strategic use of regulatory commissions. But here again, enterprisation cannot be taken for granted.

De-integration

Beyond corporatization and the setting up of regulatory commissions, the promoters of reform also argue that unbundling or de-integration of the former SEBs into generation, transmission and distribution companies is to be promoted. The aim—apart from making companies easier to privatize—is to assess the results and the responsibilities among the various activities by de-integration.[9]

But even if structures are unbundled, what really matters is that each of them becomes actually enterprised, which is not guaranteed under de-integration only.

Structural reforms undoubtedly constitute a first step towards entreprisation. However, after them, the latter is still incomplete. Privatization is one way of doing it. Public reform, if well designed and supported by some policy makers, could be another. It might be useful to discuss their respective pros and cons.

2.4 CONCLUSION—ENTERPRISING COMPLETELY: PRIVATE *vs.* PUBLIC REFORM

Both attemps at reforming, either within the public sphere or with the help of the private sector show bottlenecks. Let us focus on public reform. From the work of already existing regulatory commissions, it appears that regulation is essentially based on reduction in energy losses. Political rationality might be to accept a lower pace of non-technical loss elimination than expected by commissions, and hence lower SEB revenue, if this provides political returns. This could be strengthened by the remaining discretionary power of SEBs. There is therefore an assumption that public reform will be less efficient in tackling losses than private reform, or at least that there will be a delay, this delay being costly. As experience of successful public reform is lacking and as the contents of a credible public reform are still to be defined and proved effective, this is still a conservative assumption. So the main bottleneck of public reform is linked with the difficulty of escaping an ancient system.

On the other hand, privatization can also present difficulties. A partial entreprisation implies incomplete information about the future of the privatized organization, either because of a weak information system, or because of a weak guarantee that independent management is possible. It leads to asymmetric information between the seller and potential buyers, as well as among various potential buyers. Let us see how this affects the level of the bid: if the job expected from the private investor is to turn incompletely enterprised SEBs into firms, it has to be profitable. This profitability is the expected reward of the private investor's work. Here is the privatization deal, between a state government aiming to give off this job, and an investor aiming to have a fair return. During the bidding process, the value attached by investors to a SEB will be the sum of the discounted expected profits. Due to incomplete entreprisation, there is low information regarding the level of actual improvements to be expected in organization, management and human resources. This may lead the bidders to lower their estimation of the value of the SEB to be sold, compared with what it would have been with higher information. Indeed, all the bidders could underestimate the real value of the SEBs and make a depreciated offer to the states. This is the main bottleneck of private reform in the short term.

The systemic approach to the administrative nature of SEB management shows that classical tools of industrial economics which are linked with the study of the firm are inadequate when it comes to understanding SEBs. This approach enables one to find consistency, if not predictability in the actual running of SEBs. Consequently, it helps to evaluate what reforms are required in order to turn SEBs into firms.

NOTES

1. See for instance S.J. Coelho, *Managing Distribution Reforms*, Ministry of Power, March 1998.
2. Categorywise ratio tariff to average cost is 1.66 for industrials, 1.64 for commercials, 0.65 for residentials, 0.14 for agriculture. Several industrials turn to captive generation, with negative macroeconomic effects.
3. The official figure for total losses is 22 per cent losses over India. Some reforming states as Orissa, Haryana, and Andhra Pradesh used to declare roughly 24, 25 and 19 per cent respectively as level of losses. After the beginning of their reforms they recognized 46, 34 and 33 per cent respectively. Also Punjab is reporting 18 per cent but is said to be 10 per cent more. Delhi which used to declare 23 per cent, has now admitted losses of 43 per cent, while the real figure is said to be over 60 per cent.
4. Report no. 17234-IN, 1997, p. 4, New Delhi.
5. See S.Y. Quraishi, 'Power Sector Regulation and Restructuring: Haryana's Experience', paper presented at the International Forum on Power Sector Regulation and Restructuring, November 1998, New Delhi.
6. For instance, after an audit in 1996, APSEB reckoned that 'loss of energy due to theft was getting included in the un-metered agricultural consumption; (after the audit it was) identified to be about 600 crore Rs.'
7. This distinction does not imply any superiority of one form over the other and only reflects a difference in nature. Firms deal with productive matters and can aim at such a single monetarily expressed objective, as the effects of decisions are subject to financial evaluation. Politics, and hence administration, has by nature to deal with competing interests, that is productive, allocative, social, legal, military. Every decision has its pros and cons. By nature, there is no formal and lasting rule for trade-offs. Decision-making in administration is done through the government's discretionary power. This leads, among other consequences, to the fact that (i) trying to evaluate the efficiency of an administration through a single criterion makes no sense, and that (ii) administrations cannot be privatized. As a corollary, public power firms can exist, provided (i) public service constraints are clearly defined, and (ii) the principal objective is the minimization of the cost of supply in the long run. Briefly, the objective needs to be clear, monetarily expressed, non-contradictory, and backed by a developed information system that allows checking of results and granting of incentives.
8. World Bank Report, no. 14298-IN, Annexes 2.5 and 4.1, New Delhi.
9. Due to network externalities (Joskow and Schmalensee, 1983), cooperative coordination is necessary, and price-led market mechanisms are not enough. The former administrative coordination—earlier performed by various ministries and secretariats—will have to be replaced by grid codes and bilateral contracts.

REFERENCES

Banfield, E.C., 'Corruption as a Feature of Governmental Organization', *Journal of Law & Economics*, December 1976.

Bomsel, Olivier, 'Enjeux industriels du post-socialisme: Capitalisation et Entreprisation', *Revue d'Economie Industrielle*, no. 72, 1995.

Chakravarty, Shukhamoy, *India: The Planning Experience*, New Delhi: Oxford University Press, French translation: *La Planification du Développement. L'expérience Indienne*, Ed. De la Maison des Sciences de l'Homme, Paris, 1994.

Coelho, S.J., 'Managing Distribution Reforms', Ministry of Power, Committee on Distribution (chaired by Coelho), working document.

Hirschhausen (von), Christian, 'From Privatization to Capitalisation: Industrial Restructuring in Post-Socialist Central and Eastern Europe', in *Industrial Transformation in Europe: Process and Context*, New Delhi: Sage Publications, 1995.

———, *Du combinat socialiste à l'enterprise capitaliste—une analyse des réformes industrielles en Europe de l'Est*, Paris: L'Harmattan, 1996.

Kornai, Janos., *The Socialist System; The Political Economy of Communism*, Oxford: Clarendon Press, 1992.

Milgrom, Paul and Roberts, J., 'Bargaining Costs and the Organization of Economic Activity', in James E. Alt and K.A. Shepsle, eds., *Perspectives on Positive Political Economy*, Cambridge, MA: Cambridge University Press, 1990.

Planning Commission, Annual Report on the working of SEBs for the year 1999-2000, Government of India, 2000.

Quraishi, S.Y., 'Power Sector Regulation and Restructuring: Haryana's Experience', paper presented at the International Forum on Power Sector Regulation and Restructuring, New Delhi, 1998.

Ruet, Joël, 'Winners and Losers of the Reform of SEBs : An Organisational Study', occasional paper, New Delhi: CSH, 2001.

Salgo, Harvey, 'Power Sector Reform and the Privatization of Distribution in India', World Bank, Delhi, Internal Working Document, March 1998.

World Bank, Staff Appraisal Report, Orissa Power Sector Restructuring Project Report, no. 14298-IN, 1996.

World Bank Project Appraisal Document, Haryana Power Sector Restructuring Project Report, no. 17234-IN, 1997.

3

Whither Electricity Reforms?

GAJENDRA HALDEA

The electricity reform strategy currently being pursued in the seven states of Orissa, Haryana, Andhra Pradesh, Uttar Pradesh, Karnataka, Rajasthan and Delhi is not in the right direction. It is neither perceived as people-friendly nor is it likely to attract the much-needed investments in the power sector. The reform process seems to be working as a sedative that suppresses symptomatic pain without addressing the disease. Worse is the situation in the remaining states where reforms are yet to be initiated. For the most part, the debate on reforms is inadequate, and lacks clarity on the issues involved, leave aside their possible resolution.

The structure of the electricity industry in independent India was laid down by the Electricity (Supply) Act, 1948 that created the State Electricity Boards (SEBs). In their initial years, the SEBs performed yeomen service in carrying electric power far and wide, but over the years they have become unsustainable, thanks to their mismanagement and politicization coupled with the economic and technological developments of the past decade. It is high time that the electricity industry is denationalized, and restructured on commercial principles.

Though the legal framework was amended in 1991 and 1998 to facilitate private investment in generation and transmission respectively, it enabled private entities to sell or transmit power only through long-term contracts with state-owned entities. Such contract-driven privatization through state-owned monopolies can have little chance of enduring success. Similarly, the setting up of regulatory commissions under the 1998 Act, though a welcome move, can only have a limited impact on the state-owned monopolies. It should surprise no one that these piecemeal changes in the name of reform have not been able to arrest the deterioration of this industry.

To overcome the outdated structure stipulated in the central laws, the seven 'reforming' states have enacted their own laws and adopted the so-called Orissa model for unbundling their monolithic SEBs into generating, transmission and distribution companies, to be regulated by an independent regulatory commission in each state. The argument in favour of unbundling was the pressing need for creation of viable commercial entities that would lend themselves to efficiency improvements and privatization, as in the West.

The similarity of the Orissa model with the restructuring undertaken in other countries virtually ends with the unbundling of SEBs. Unlike the West where competition in generation and supply is the engine for efficiency gains and tariff reduction, the Orissa model relies on an interconnected chain of monopolies where competition is conspicuous by its absence. In effect, it has only tinkered with the structure without incorporating any sound economics of regulation.

3.1 SINGLE BUYER MODEL

Orissa represents a 'single buyer model' where all generating companies (GENCOs) are required to sell their produce to a state-owned transmission company (TRANSCO). This implies that even if GENCOs are willing to offer spot sales, or enter into short-term contracts, there can't be a credible market in the absence of multiple buyers. Gencos cannot, therefore, bear the market risk and must rely on long-term power purchase agreements (PPAs) with the TRANSCO. As a result, new capacity can only be contracted through the state-owned TRANSCO that continues, like its predecessor the SEB, to negotiate PPAs on a 'cost plus' basis leading to comparatively high tariffs.

In the Orissa model, the absence of competition does not stop at generation. Since GENCOs must sell their produce to the TRANSCO, distribution companies (DISCOMs) can buy from the Transco alone, and the consumer must, in turn, source all his requirements from the DISCOM of his area. The industry structure thus continues to be in the command-and-control mode, unencumbered by competition and consumer choice. In this chain of monopolies, public as well as private, all prices are determined on a 'cost plus' basis either through negotiations or by the regulator, and this constitutes a perfect recipe for delivering high cost power to the consumer.

At the distribution end of the Orissa model, the prevailing tariff structure coupled with the high transmission and distribution (T&D) losses do not permit adequate cost recovery by DISCOMs who persistently default on their payment obligations to the state-owned TRANSCO. The use of TRANSCO as a free banker for DISCOMs, carrying their unpaid bills aggregating about Rs. 900 crore, negates the very rationale of commercialization and privatization. Privatized DISCOMS in Orissa are virtually being allowed to accumulate unsustainable losses year after year without any clear road map as to when, if at all, these losses will be wiped out, and by whom.

The losses/receivables of DISCOMs in Orissa are currently being accumulated on the balance sheet of the TRANSCO whose outstanding payments to the state-owned GENCOS continue to rise. With a negative net worth, the TRANSCO may be able to contract new generation capacity, if at all, only on the strength of individual state guarantees and escrow arrangements that are widely regarded as unsustainable. In practice, therefore, Orissa as well as the other 'reforming' states may fail to attract much capacity addition and the prevailing power shortages would continue to persist.

The TRANSCO in Orissa has already accumulated overdue payables of more than Rs. 1,500 crore that continue to mount steadily, besides a debt burden of about Rs. 3,000 crore. It seems to be caught in a debt trap that can only be addressed through steep tariff increases, which are unlikely to materialize. The state government may, therefore, have to rescue the TRANSCO—a doubtful proposition in view of the precarious condition of state finances—or else this rising burden may stunt the performance and growth of the industry. Evidently, the Orissa model in its present form is not sustainable. This model was perhaps adopted as a first step towards restructuring, but the need to move forward is now imminent.

It appears that just as a reform package had to be evolved for dealing with the financial chaos that was the SEB, another package may be necessary for bailing out the DISCOMs and the TRANSCO from their impending bankruptcy. The added complication, however, would be the legal and moral hazards posed by any state assistance to private DISCOMs for cleaning up their losses. Similar problems may arise if DISCOMs are allowed to charge higher tariffs from future consumers for offsetting their past losses that were disallowed by the regulator.

3.2 BANE OF LONG-TERM PPAs

As a matter of principle, creation of monopolies must be regarded as an unacceptable form of interference by the government in the operation of free markets. It is pertinent to note that in several developed countries, induction of competition in telecom and power has resulted in significant tariff reductions during the past decade, but these lessons seem to have been completely ignored in India insofar as power sector reforms are concerned. While competition in telecom is beginning to show dramatic results, there is as of yet no move towards introducing any worthwhile competition in the electricity industry.

It is important to recognize that in the emerging structure of the electricity industry across the world, long-term PPAs are no longer regarded as conducive to consumer interests. In India too, there are a number of PPA-based projects where the capital costs and operating norms, as approved by the Central Electricity Authority (CEA) and the respective state governments, leave much room for improvement. These PPAs have typically allowed select private entities to procure power plants based on indicative capital costs approved by CEA even though similar approvals for public sector entities like the SEBs and the National Thermal Power Corporation mandate open competitive bidding. In a 'cost plus' regime where capital costs directly translate into tariffs, this issue deserves a more credible treatment. Compelling the consumer to buy power for 20 years or more from a source that has been determined through a sub-optimal process sans competition can hardly be regarded as a practice worthy of emulation.

In the UK, for instance, GENCOs do not have the comfort of long-term PPAs to assure them of guaranteed sales and profits. Until recently, they were required to sell their produce through a 'pool' where half-hourly bidding determined the price and offtake. With a view to eliminating the possible manipulation of pool prices through the use of market power by some GENCOs, UK has recently amended its laws to facilitate licensees/consumers to buy directly from competing GENCOs. In some of the other developed countries, existing PPAs have been terminated, with compensation, to allow for competitive sale of power in the market. Thus the typical PPA seems to be facing extinction in the developed markets; India can hardly hope to usher successful reforms by using the PPA as a vehicle for capacity addition.

In the reform strategy, the central objective should be the creation of a sound structure that promotes growth coupled with efficiency. Towards this end, generation and supply should be separated from transmission with a view to subjecting the former to competition. The proposed separation between the carriage and content businesses is somewhat similar to telecom where several long-distance or cellular operators compete even where a single wire connects the consumer's telephone. Since generation typically constitutes over 75 per cent of the consumer tariff, competition among GENCOs selling to multiple buyers would bring about significant price and efficiency gains, whereas overlooking this fundamental economic principle would be tantamount to ignoring consumer welfare in the reform design.

In the restructured industry, the TRANSCO must not buy or sell power; it should only transmit on payment of regulated wheeling charges with a view to providing open access to its transmission network. At the distribution end, DISCOMs must provide similar open access to their wires for enabling bulk consumers to buy directly from GENCOs. With these fundamental elements of restructuring in place, market forces can be relied upon to bring in rapid efficiency gains as well as new investments that have so far been elusive. It is important to recognize that an efficient and competitive industry is a prerequisite for achieving the goal of universal access.

The industry structure proposed here is predicated on the provision of non-discriminatory open access to the 'transmission highways'. A TRANSCO that buys and sells electricity would face a conflict of interest in transmitting the electricity owned by its competitors, and it is for this reason that companies operating transmission networks in the restructured markets abroad are prohibited from buying and selling power. This is critical for creating a free market where GENCOs and suppliers can use a common network for selling directly to bulk consumers and DISCOMs. Besides providing a competitive environment, this arrangement would enable GENCOs to assess market demand, enter into contracts with bulk consumers, and set up 'merchant power plants' without relying on government guarantees. The creation of such capacity will free some of the existing supplies of bulk consumers that would become available for the benefit of other consumers.

3.3 REFORM OF DISTRIBUTION CRITICAL

There is increasing recognition that reforms at the distribution end are critical for restoring the viability of this industry. This would require distribution to be depoliticized and privatized with a view to containing the large-scale collusive thefts and for upgrading the network. Though difficult to achieve, sustainable privatization of distribution presupposes a well designed and comprehensive regulatory framework that is transparent and predictable, unlike the present 'cost plus' year-to-year approach that does not sufficiently incentivize efficiency improvements, and on the contrary fosters regulatory uncertainty.

Complete separation between the wire and content businesses at the distribution level may not be feasible without a substantial systems upgradation that is unlikely to be cost-effective at this stage. Yet, open access to the wires of a DISCOM must soon be introduced in respect of bulk consumers. Since these consumers possess the requisite bargaining strength and tend to consume large quantities, they can buy directly from competing producers and bring about the desired competition in the entire industry. The introduction of third party access to the distribution networks would thus usher the much-needed competition in generation and supply, besides providing a credible mechanism for adding generating capacity.

For ceding their bulk consumers to GENCOs, the DISCOMs can be suitably compensated through the levy of a wheeling surcharge or electricity duty that would help in sustaining the subsidized farm tariffs. Cross-subsidies are here to stay, and competition cannot be postponed indefinitely until their elimination. The problem can, however, be tackled by simply isolating the cross-subsidies and collecting an equivalent surcharge or duty from bulk consumers. As long as they pay such a surcharge or duty, bulk consumers should have the freedom to buy from competing producers.

In the structure proposed here, continued regulation of tariffs payable by retail consumers, not being bulk consumers, would be necessary for guarding against volatility and speculative pricing. Adequate supplies would also have to be ensured by earmarking the existing sources of cheaper generation for their consumption. In effect, this would imply that the existing sources, exceeding 1,00,000 MW of generating capacity, would not enter the competitive market, and would continue their supplies as per extant agreements. Only new capacity would be subjected to competitive markets, to begin with; and competition would thus be introduced at the margins. States would be free to subject the older power stations to a similar regime as the markets mature. Such an approach would ensure a smooth transition for the markets as well as the consumers.

It is nobody's case that subsidies should be eliminated; on the other hand, there is an urgent need to target them towards well-defined recipients. Affirmative action by the government would continue to be necessary for providing universal access at affordable prices, particularly in rural areas and for the economically weaker sections. But decades of experience shows that governments alone, howsoever resourceful and well meaning, cannot accomplish the task without the market. In fact, the poor would be better served by market-driven policies, topped with targeted subsidies.

While addressing the problems arising out of near-free supplies to the farm sector, it is important to recognize that cross-subsidies have pushed up industrial tariffs to almost double the household tariffs in India whereas it is the other way round in several developed countries. If Indian industry has to face global competition, the quality and price of electricity would matter, and a sustainable approach to subsidies would be necessary. The surcharge proposed here would not only help in funding the subsidies, it would also ensure transparency in determining the extent of such subsidies as well as the government subventions necessary for this purpose.

3.4 PILFERAGE: THE REAL PROBLEM

Commercial accounting in the 'reforming' states is beginning to reveal the cesspool that the electricity industry seems to have become—how else does one describe a bizarre situation where more than 40 per cent of the produce does not get accounted or paid for? Pre-reform, the T&D losses were usually stated in the region of about 20 per cent, with poor systems maintenance and outlying rural areas being held as the main culprits. Pilferage by small and large consumers alike, apparently in collusion with the SEB employees, now seems to be emerging as the real problem. This is a shocking state of affairs as no civil society can afford to overlook such widespread prevalence of theft for which honest citizens are penalized, or worse, the burden is shifted to future generations by renaming these losses as 'regulatory assets' that would qualify for tariff increases in subsequent years. This failure of governance needs to be rectified through stringent and effective measures lest the judicial courts intervene, as in the Delhi pollution case, and take charge owing to the government's abdication of its legitimate role.

In the debate on reform, there has been a sharp focus on the low tariffs that do not allow for cost recovery. This argument is being used for strengthening the case for tariff increases. There is, however, inadequate focus on pilferage and technical losses that account for over 40 per cent of the electricity generated. For example, Orissa continues to report T&D losses of about 43 per cent while in Delhi they reportedly exceed 50 per cent. The prevailing approach that relies on a 'cost plus' determination of tariffs on a year-to-year basis is not conducive to the reduction of these losses at a pace sufficient to restore the viability of the industry. A regulatory framework that incentivizes loss reduction in a transparent and predictable manner is, therefore, essential. In addition, if competition were introduced in generation and supply of electricity, it is bound to push the tariffs downwards. As such, while some increases in tariff may be justified for ensuring cost recovery, these should be largely offset by reduction in T&D losses on the one hand and introduction of competition on the other.

Whether India is able to realize the full potential of reforms depends crucially on the government and the regulators whose role is vitally important. As California's example has shown, governments can make a big difference by getting it wrong, often under pressure from the industry's incumbents. In fact, that has been the sad story of India's power sector during the past decade. As for the regulators, they should be willing to trust market forces; they must make the rules of the game clear and refrain from arbitrary interference during the transition; they must detect and curb market abuses effectively; and they should ultimately yield most of their powers to the market.

3.5 CONCLUSION: THE ELECTRICITY BILL

In sum, the industry structure being advocated here would help introduce competition, improve efficiency, add capacity, rationalize tariffs and enhance consumer welfare. It would unleash latent energies, enterprise and innovation that should galvanize the industry towards rapid growth and enable consumers to enjoy lower prices with improved services. On the contrary, continued adherence to the Orissa model in its present form would promote monopolies, raise tariffs, deny consumer choice and constrain investments in the 'reforming' states. Reforms would not be perceived as people-friendly if tariffs rise without a perceptible improvement in the quality of supply. As for the remaining states, failure to initiate reforms expeditiously would lead to a virtual collapse of their power sector as well as the state's finances.

During the past decade of economic liberalization in India, electricity has perhaps been the only major industry that has failed to improve even though it is so vital to economic growth and human development. It has remained comparatively static in the midst of significant changes that are taking place elsewhere in the world. Several principles and practices that have universal relevance are yet to be applied to the power sector in India. If a GDP growth of 8-9 per cent is to be achieved, there is no way it can be done without a holistic restructuring of the power sector, and the proposed Electricity Bill should form the vehicle for this inevitable change.

PART II

TOOLS: BETWEEN REFORMING AND PRIVATIZING SEBs

4

Internal Audit as a Reform Technique for Power Sector in India

R.K. MISHRA

The power sector is one of the most capital-intensive industries in India. Power costs are a dominant cost component of any economic activity.[1] Recent statistics reveal that commercial losses of State Electricity Boards (SEBs) have exceeded Rs. 20,220 crore during 2000-1 with APSEB/APTRANSCO topping the list with estimated losses of about Rs. 3,800 crore. 19 SEBs across the country are estimated to have netted Rs. 82,462 crore worth of revenue during 2000-1 and reported transmission and distribution (T&D) losses of 23 per cent.[2] The mismatch between cost of power and tariff is cited as the main cause for such losses.

The tendency towards competitive rates and the need for attaining increased productivity propelled by reforms in the Indian power sector indicate the underlying paradigm shift taking place. Reforms can be twofold: internal and external. The power sector reforms in India are based on externalities as propounded by the World Bank. The history of successful reforms shows that internal measures are first initiated prior to the application of external initiatives. Of course, there may be certain measures that could qualify as both internal and external.

The Optimum Performance Through Internal Management Action (OPTIMA), advocates the application of internal measures as a technique to reform an enterprise. OPTIMA is an evolving approach which encompasses tools such as organizational restructuring, financial restructuring, product portfolio restructuring, labour restructuring and environmental interface restructuring. It is preferable to turn an enterprise around through internal restructuring than take recourse, in the first instance, to external restructuring. Internal restructuring is comparatively more economical, more effective, speedier, durable and serves as a good accomplice to external restructuring. The present paper focuses on the introduction of effective internal audit in the power sector in India which is an important component of OPTIMA.[3]

The chapter is based on research carried out in the case of nine power utilities in the state sector sponsored by the Power Finance Corporation of India (PFC) at the initiative of the World Bank.[4] The chapter provides the backdrop, scope, charter, coverage, reporting arrangements and performance of the internal audit in these utilities. The paper also examines the adequacy of internal audit unit in terms of staffing, qualifications and experience of internal auditors and the need for training of the internal audit staff.

4.1 BACKDROP

To begin with, let us examine the background of the Internal Audit (IA) function in the various power utilities. Table 1 below presents the strength and total expenditure on IA in the different utilities.

The Andhra Pradesh State Electricity Board (APSEB)/APTRANSCO, had a gross annual expenditure of Rs.1,600 crore per annum. Its gross annual revenue was of the same order. It employed more than 70,000 employees. Its budgeted staffing level for the Internal Audit Department (IAD) was 91. Its annual IAD staffing level during the same year was 66. Its annual expenditure on IA was of the order of Rs. 54 lakh per annum.

TABLE 4.1: STRENGTH OF AND EXPENDITURE ON IA IN DIFFERENT UTILITIES

Name of the Board	IAD Strength	Expenditure (Rs. in lakh)
APSEB	66	54
ASEB	39	26
KEB	421	48
KPC	17	3
KSEB	221	NA
MSEB	103	60
NEEPC	9	5
RSEB	315	215
UPSEB	192	175

The Assam State Electricity Board (ASEB) incurred a gross annual expenditure of Rs. 305 crore. Its annual revenues were about Rs.140 crore. It employed 23,000 persons. The IAD staff in the Board totals 39. The total expenditure on IAD was Rs. 26 lakh per annum.

The Karnataka Power Corporation (KPC) had 17 persons in its IAD. It was engaged only in the transmission of power. This is because in Karnataka, transmission and generation of power had been separated, and consequently two organizations were dealing with each of these functions. It had 422 employees in its IAD.

In the Maharashtra State Electricity Board (MSEB), the gross annual expenditure was about Rs. 3,000 crore. The Board received an equal amount of revenue to match its annual expenditure. It had on its rolls about 1,20,000 officers and workers. It had about 103 persons in its IAD. The total expenditure in IAD was Rs. 60 lakh per annum.

The North Eastern Electric Power Corporation (NEEPC) is a central undertaking. Its gross annual expenditure was about Rs. 35 crore and gross annual revenue was Rs. 49 core. It had a strength of about 4,000. It had only 9 persons in its IAD. The expenditure on maintenance of the IAD was about Rs. 5 lakh per annum. The Uttar Pradesh State Electricity Board (UPSEB), during the same period, incurred an annual gross expenditure of Rs. 175 lakh. It had 192 officers in its IAD.

A brief examination of the general background of IA in the SEBs reveals that IA as a function has already made its presence felt in these boards. The percentage of the expenses incurred on IA, and the number of employees working in the IAD as a percentage of gross annual expenditure and total number of employees, differs from board to board. During the discussions with officials of some SEBs, it was indicated that in view of the increasing importance of IA in their operations, they propose to allocate higher outlays for their IADs to enable them to increase the level of staffing and provide higher sums for travel, etc.

4.2 SCOPE AND COVERAGE OF INTERNAL AUDIT FUNCTIONS

4.2.1 Scope

The scope of IA function, although having some commonalties, differs very widely in a couple of aspects in respect of the boards under study. In the case of the APSEB, the scope of the IAD covers aspects such as ensuring compliance with laws and regulations, administrative policies, plans and procedures, ensuring soundness and application of management controls over administration, revenues and expenditure, ascertaining efficiency in acquisition and use of resources, checking upon the effectiveness in achieving organizational objectives, and ensuring effectiveness in achieving the overall programme of the Board. IAD uses both systems and transactions-based audit to accomplish these tasks. The activities open for the scrutiny of IA, include security classified information, records protected by piracy provisions, and important policy areas. About 10-15 per cent of the APSEB's activities get covered by the IA. The IAD is often involved in the preparation and review of financial statements. It also attends to activities pertaining to the control of frauds including fraud detection and fraud investigation. The IAD is not consulted in the design and development stage of new or revised programmes and systems.

The ASEB also includes in the gamut of IA, activities such as revenue audit, expenditure audit, stores audit, and special audit relating to any or all of these activities. The Board considered the expenses incurred on IA as worthwhile as it derives full value for money expended in IAD. Among the benefits accruing from IA activities to the Board, the prominent ones are savings and economies on the various heads of expenditure, improved control, and prevention of shortfalls in respect of various items of operation. The Board is not fully satisfied with the performance of IA, and the main reason for this is the inexperience of the IA staff.

KPC has the following objectives for its IAD:

1. Reviewing the soundness, adequacy and application of accounting, financial, and operational controls.
2. Examining and reporting on the extent of prescribed procedures under rules/statutory requirements in regard to financial and accounting transactions.
3. Reviewing and reporting the action taken by the departmental authorities on the points brought out in previous audit reports, both internal and external.

KSEB includes under the ambit of IA activities, revenue audit, cash audit, expenditure audit, test audit of stores and pre-audit of project-related purchases. Before undertaking IA the IAD formulates a plan of action. KSEB feels that IA provides value for money spent on its activities. The Board is also of the view that IA activities have benefited the organization in several ways in recent years—by affecting economies, improving control, and ensuring justification of additional resources required by its various departments and systems and programmes. The Board is satisfied with the standard of performance achieved by the IAD.

Under the gamut of its IA activities, MSEB includes, verification of stores, and revenue and establishment expenditure. It plans IA activities on a half-yearly basis. The Board feels that the IA activities provides value for money spent. It is of the view that the introduction of IA has ensured that the different functions of the Board are discharged more efficiently. It has resulted in increased management reassurance, provided superior justification for reallocation of resources and affected savings and economies in expenditure. Though the Board is on the whole satisfied with the functioning of the IA, it is of the view that the audit personnel's lack of experience has prevented the realization of the full benefits of this function to the organization. The Board also feels that the inordinate delay in complying with IA objectives by the department does not allow it to do justice to its other activities.

In the case of the NEEPC, all the activities except the ones which require management decision, are included in the purview of IA. The corporation plans its IA activities from year to year. It feels that the investment in IA has proved beneficial. Among the benefits of IA accruing to the organization, the prominent ones are: identification of significant shortfalls in compliance, efficiency and effectiveness, and improved control of procedural arrangements.

4.2.2 Audit Charter

The IA Charter, also known as the IA Manual, is a document specifying the role, responsibilities and authority of the IA function in an organization. All well-organized enterprises put such a charter in place before commencement, or in the process of initiating their operations. The SEBs are very vast organizations by Indian standards. Their working is complex as compared to a typical commercial organization, as they have to perform the dual act of providing a service on a commercial basis as well as fulfilling their social obligations. In such a situation, the presence of an IA Charter assumes greater significance. The KPC, and MSEB have prepared their own IA Manuals. The rest of the SEBs covered by our study, including APSEB, though recognizing the need for a charter, have not been able to develop one. They also mentioned that they did not have any plan to prepare an IA Manual. Two of the corporations having an IA Charter disclosed that though there is a need to revise it within a period 1-2 years, it took them much longer: the KPC revised the IA Manual 3-5 years ago, while the KEB revised 6-10 years ago. The MSEB was an exception, the IA Manual having been revised within a period of 12 months. However, the SEBs have definite views about the highest level of endorsement of the IA charter. In the case of APSEB and KEB, it was suggested that the Member Finance should endorse the IA charter. In the case of ASEB, KPC, and MSEB, the competent authority suggested for endorsement was chairman/chairman-cum-managing director. A review of the constraints imposed by the charter on the scope of IA activities on the various SEB points out that no areas were omitted from the gamut of IA.

4.2.3 Audit Committee

An Audit Committee normally comprises the members of senior management of an enterprise who oversee and review the operations of IA functioning. In enterprises of world fame, a special role is assigned to such companies. For instance, Unit Trust of India has such a committee, as also ACC Limited. Such committees are necessary in the case of huge enterprises in terms of operations and investment, requiring these to have an effective view of operations and to undertake an examination to ascertain the effectiveness of the control systems. These committees are normally set up by the Board and may comprise members drawn from the Accounting, Economics, and Law disciplines. The committee might also have experts with an intimate knowledge of the functioning of the industry of which the firm is a constituent. The recognition of the need to bring the IA function under the control of the audit committee implies the amount of importance that an enterprise attaches to the audit function. Our discussions with the concerned authorities in the various SEBs reveal that no board deemed it fit to set up an audit committee to oversee the functioning of IA.

4.2.4 Internal Audit Plan

The presence of a strategic audit plan is a primary requirement for the success of the IA function in an organization. Such a plan should spell out the duration, audit priorities, and the target maximum audit cycle. In the case of the APSEB, the strategic audit plan had a target maximum audit cycle of one year. While formulating such a plan, no consideration was given to APSEB's other strategic plans. It may be noted that no audit priorities were fixed in the case of the APSEB and, therefore, no question arose concerning the importance of various factors which have to be considered for fixing such priorities. The ASEB had not given any thought to the question of formulation of such a plan. In the case of the KPC, the strategic audit plan had 1-2 years duration, and no reference was made to the organization's other strategic plans. While formulating the audit plan, priorities are fixed by giving maximum importance to factors such as materiality, risk, impact on other operations, public interest, recent review activity, and audit resource requirements. In the case of the KEB, the strategic audit plan had a duration of 1-2 years. Audit priorities were fixed giving due regard to various factors as mentioned in the case of the KPC. MSEB did not have a strategic audit plan. However, it regarded materiality and public interest as very important, and impact on other operations fairly important, which were to be given due weightage in the exercise of framing any such plan. The Board suggested a target maximum cycle of two years.

NEEPC did not formulate a strategic audit plan. However, it suggested a target maximum cycle of two years and indicated a three-year maximum cycle based on its current experience.

4.2.5 Internal Audit Coverage

The IA coverage determines the success or failure of the IA function in any enterprise. A big gap is found between the targets and achievements in this area. The question of coverage is shrouded in mystery: '*should it be only the coverage of operations or should it also span the functioning of the various controls?*' Some experts have also raised the issue of horizontal interface while talking about the coverage of IA, thereby making the suggestion that a firm's efficiency should be compared with similar firms in the same industry. Our discussions with the officials of the various SEBs pointed out to unanimity only with regard to the operational aspect of IA coverage. In the case of the APSEB, the audit coverage fluctuated between 50-60 per cent of operations, as against 90-9 per cent in the case of the KPC, the KEB, and the MSEB, 80 per cent in the case of NEEPC, and 40 per cent in the case of ASEB. The significant reasons for not completing the planned IA coverage included shortage of IA staff, staff inexperience, underestimation of the time required to complete the planned audits, new responsibilities assigned to IA function, major administrative changes in auditable areas, IA overtaken by other reviews, etc. The strategy suggested to address the lack of coverage of auditable areas was clearing the backlog of inspections in the subsequent programmes by the APSEB. In the case of the KEB a consultant was appointed to suggest ways and means to remove the backlog of IA. In the NEEPC steps were taken to assess time requirements more rationally. In the case of the MSEB action was taken to redress the problem by initiating training for the officers and staff. The IAD was allocated additional staff. In the ASEB, though action was taken to clear the backlog, the problem was only partially solved. Some extra posts were created in the IAD, selections made, and postings done. In the case of the APSEB, action was being considered to increase the audit coverage, and continually audit the activities brought under such coverage. The problem of shortage of staff was temporarily handled by directing the circle superintending engineers to draft Junior Accounts Officers for IA on a deputation basis in the absence of experienced Junior Accounts Officers at the headquarters.

IA comprises audit of operations concerning finance and accounting, administration and management, operations and services, and allied matters. In any approach to undertaking IA operations, an optimum balance should be struck with regard to the time allocated for IA of these various activities. A study of the time allocation reveals that in the absence of any sensitization to this important aspect, the APSEB, the KPC, the KEB, the MSEB, and the ASEB have not allocated any specific time. Only the NEEPC indicated that 66 per cent time was allocated to IA for the financial and accounting operations, with the rest being allotted to the other operations of the organization. A stock taking of the purpose of IA was restricted to dimensions such as the integrity of information, compliance with laws and regulations, administrative policies, plans and procedures, soundness and application of management controls over assets, revenue and expenditure, economy and efficiency in acquisition and use of resources, effectiveness in achieving organizational objectives, and effectiveness in achieving programme objectives. KPC and KEB wanted to achieve all the above-mentioned purposes of IA. The NEEPC had its focus on integrity of information and compliance with laws and regulations, administrative policies, plans and procedures. The other purposes of IA were considered less important. The APSEB indicated that it considered soundness and application of management controls over assets, revenue and expenditure, as the driving force of the IA function.

One of the important issues in IA is how to evaluate the evaluators themselves. When this question was posed to the various SEBs interesting revelations were made. In the case of the APSEB, the performance measures used to judge the efficiency and effectiveness of the IA function were indicated as the review of monthly returns submitted by IA units, realization of shortages, for which replies were received from field offices. The KPC, the KEB, the NEEPC, and the ASEB realized the need for having such a performance indicator, but they were yet to develop one.

4.3 REPORTING, ORGANIZATION AND TRAINING

4.3.1 Reporting Arrangements

The success of any IA system depends on how the arrangements for reporting the various findings are put in place. The practices in this connection reveal that in the case of the APSEB, audit findings are documented and communicated to operational management, an executive summary of an audit report is provided to the user departments; and audit recommendations are always followed up by an IA. However, the operational management does not always have to respond to the IA findings and recommendations. The proportion of audit recommendations made in IA reports which were implemented ranged between 50-60 per cent. About 15-25 per cent of the recommendations, though accepted, were not implemented. The main reason for this non-implementation was the lack of resources available with the APSEB. In the case of the KPC, IA reports are submitted to the chairman-cum-managing director as opposed to the operational management in the case of APSEB. The operational management is expected to respond to audit findings and recommendations in a timely manner, and in practice the operational management complies with these requirements. In the case of the KEB, IA reports are submitted to the chairman, member, finance department, and concerned administrative officers. In the case of the NEEPC, the IA reports are submitted to the operational management and the external auditors. The audit findings are documented and communicated to the operational management and followed up by IA. The operational management is also required to respond to audit findings and recommendations in a timely manner. About 50 per cent of the recommendations made by IA were accepted. Again, about 15-25 per cent of the recommendations, though accepted, were not implemented. A large number of recommendations were not even considered. In the case of MSEB, the IA reports are submitted to the operational management. The audit findings are documented and communicated to operational management, and the audit recommendations followed up by IA. The management is required to always respond to such audit findings and recommendations. In the case of the ASEB, the IA reports are issued to member, finance department, and operational management. The audit findings are documented and communicated to operational management and such recommendations are followed up by IA. However, the operational management normally does not respond to the audit findings and recommendations in a timely manner.

4.3.2 Use of Contract Internal Auditors

At a time when privatization of the activities of the SEBs is gaining ground in the country, it is worthwhile to consider whether the IA function could be contracted out to produce better results. All the SEBs studied by us responded in the negative when the idea of contracting out the IA function as a whole, or some of its elements, was put forward to them.

4.3.3 Internal Audit Unit

All the SEBs under the study had their own IA unit. APSEB had a staff of 66 persons and an expenditure of Rs. 54 lakh. It spent Rs. 45 lakh on salaries and Rs. 9 lakh on administrative and operating expenses. The IA unit of the organization had no plans to recruit and retain suitable staff. The total number of staff employed by KPC was 17, comprising 6 posted in the central office, and 11 in its various regions. The staff are interchangeable between the accounts and other departments. In the case of the KEB also, the staff of the IAD was interchangeable with the accounts and other departments. The total staff of 422 persons in IAD comprised 15 posted in the central office, and 407 in the various regions. In the case of NEEPC, there were 9 officials working in IAD. The total cost of running the Department was made up of Rs. 4,08,000 as salaries and allowances, and Rs. 50,000 as administrative and operating expenses. All the 9 functionaries were posted in the central office. In the case of MSEB, the total staff of 103 persons in IAD comprised 46 working in the central office and 57 posted in the various regions of the Board. The total expenses incurred on IA were Rs. 60 lakh, made up of Rs. 49 lakh towards salaries and allowances, and Rs. 11 lakh towards administrative and operating expenses. ASEB employed 39 persons in its IAD, and all of them were based at the central office. The yearly expense incurred was Rs. 26.40 lakh, of which Rs. 18 lakh was spent on salaries and allowances, and Rs. 8.40 lakh on administrative and operating expenses.

4.3.4 Qualifications and Experience of Internal Auditors

There is a trend today to staff the IAD with well-qualified and experienced personnel. Most of the organizations recruit professionals and Internal Auditors and pay them very well. In the case of the APSEB, most of the IAD staff have only a Bachelor's degree. In the case of the KPC, excepting one executive who held a CA/ICWA degree, the others were graduates. In the case of the NEEPC, of the 9 executives employed in IAD, there were 2 chartered cost accountants, 1 ACS, 1 M.Com/MBA, 1 Law graduate, and 5 holding a Bachelor's degree. Some professionals were of more than one professional body. In the case of MSEB, there were 2 chartered cost accountants, 2 M.Com/MBA, and 1 Law graduate, with all the others holding a Bachelor's degree.

2 executives were members of more than one professional body. In the case of ASEB, excepting two persons who had a Master's degree in Commerce, the others were all graduates.

4.3.4 Training of IA Staff

The experts in IA have suggested continuous training of the IA staff. The APSEB neither has any ongoing formal training for its staff nor any plan to launch such training programmes, though during the course of informal discussions the need for such training was emphasized more than once. The state of affairs in the ASEB, the KPC, and the NEEPC was similar. The KEB and the MSEB had an ongoing formal training programme. An officer was assigned the task of planning and developing training programmes and organizing such programmes with the help of internal resources and external agencies. MSEB trod the same line. The IAD was fairly satisfied with the coverage of training for its staff, the training areas, the number of staff receiving training, the content of training courses, and the presentation of training courses. However, it felt that the need for training existed and it did not receive any worthwhile attention, including system-based auditing, basic performance auditing, and advanced performance auditing.

4.3.6 Authority of Chief Internal Auditor

A survey of the authority of the chief internal auditor in the various SEBs showed that in APSEB the head of the IA function was authorized to develop audit priorities, finalize work plans and programmes, and determine boundaries and the direction of audits. However, the Department did not divulge as to whether they had direct access to those who oversaw and reviewed the IA function. The chief of the IA function also controlled the region-based IA units. The case of the KPC was no different. In the case of the NEEPC, the chief of IA had the authority to develop audit priorities, finalize work plans, and chalk out programmes, but he had no authority to determine boundaries and direction of audits. He did not have any direct access to those who oversaw and reviewed the IA function. In the case of the MSEB, the chief functionary of IAD was authorized to set out audit priorities, finalize work plans and programmes, and determine boundaries and direction of audits. He also had direct access to those who were appraising the performance of the IA function. He exercised control over the region-based units. The case of the ASEB was similar, except for the fact that the chief of the IA function did not have access to those who oversaw or reviewed the functioning of IA.

4.4 IA PERFORMANCE

Under this head, we propose to study IA performance in the various SEBs at the chief executive, head of the department, and operating manager's levels.

4.4.1 ASEB

A full-fledged IAD began functioning in 1975. It was subsequently upgraded and is now headed by the chief internal auditor, equivalent to the rank of chief engineer. He is assisted by the deputy chief internal auditor. The status of the chief internal auditor is equal to the head of any other department. He reports to the chief of the finance department. The IAD has also been strengthened and nearly six teams are simultaneously engaged in field audit. Traditional IA is conducted at least once in two years in the respective units of the organization.

4.4.2 KEB

The IAD in KEB was set up in 1957. There were 221 persons on the staff of IA. It was headed by the financial adviser and chief accounts officer, who was a B.Com and FCA, and senior accounts officers, who were SSLC matriculates. They all had at least two to three decades of experience in their job.

The chairman accepted the importance of IA as an essential instrument of financial control. He agreed that IA serves as a powerful arm to the management in appraising the activities of the organization. He favoured having the chief of IA as part of his management team, but disclosed that the IA staff was not very capable of thinking in managerial terms and did not posses the necessary technical skills. He wanted IA to be viewed as an independent appraisal activity and recommended a certain amount of independence in its functioning. He wanted highly qualified people to staff the IA and the IA team to be interdisciplinary, reporting only to the managing director. He wanted the IAD to conduct surprise checks and IA to be conducted on a continuous basis. He did not want IA to be burdened with additional responsibilities. He pointed out that IA was manned by under-qualified people and lamented their lack of training. He expressed unhappiness about the practice of frequent interdepartmental transfers.

The chief of the IA function complained about the lack of adequate strength in the IAD. KEB's IAD was involved in appraising its activities in terms of policies and procedures. However, it was not undertaking physical verification of assets. He admitted that the effectiveness of the team was limited to the extent of its having on the staff only accountants. In his view, the work done by the

IAD was not helpful in taking managerial decisions. On the other hand, he was satisfied with the work done by the IAD in connection with bringing out cases of defalcation and defraud. While he said that the employees of his department, including himself, were happy with their status, they were only happy to some extent with job content and power to perform their work. He disclosed that the Department was going by test checking only. His audit staff while conducting the audit employed techniques such as physical examination, confirmation, searching, examination of documents, and correlation with related information. However, he accepted that his department was not conducting systems audit and was restricting itself to traditional audit. On being asked whether he was getting cooperation from the other departments while conducting audit, the answer was 'only to some extent' as far as Purchase, Stores, Services, Maintenance and Administration departments were concerned. He was not getting support from the production, personnel, and research and development departments. He also did not invite suggestions from other operating heads while planning IA.

The operating managers in KEB held that the role of IA should be that of an expert reviewer of the operations. It should suggest measures to improve the operational efficiency. They suggested an effective IA mechanism to make the organization perform well. They also recommended granting of adequate freedom to the IAD. In their view, internal control measures were a precondition to the success of IA. IA should be given sufficient independence, but not vigilance powers. In their view, IA should be reporting to the managing director and staffed with technically qualified people having knowledge of operations. The audit team should have representation from each department of the organization. IA is considered essential for an organization to evaluate the efficiency of the management system. The reports submitted by the IAD should contain the comments of the respective heads of department against the remarks made by IA. IA should be empowered to undertake operation audits. The technical competence of the IAD should be able to match this challenge. IA staff should be exposed to practical working in operations. They pointed out that the management hardly ever utilized the IAD expertise. This could be because the persons working in the IAD were not competent, which ultimately led to IA not exercising effective control in the organization. They expressed skepticism about the role of IA in contributing to general effectiveness in their operations.

4.4.3 MSEB

The Chairman of the Board agreed that an effective system of IA was an important instrument of financial control, and that IA could serve as a powerful arm of the management. He suggested management training for the IA staff to enable them to come up to expectations in this regard. He expressed his unhappiness about the lack of qualified staff in the organization, and felt that this deficiency could be removed through training. In his view, the chief of IA need not necessarily possess any professional degree, but what was more important was his attitude to the job, besides his professional background. He wanted IAD to be staffed by a multi-disciplinary team and have a reporting relationship with the board audit committee. While he suggested surprise checks by the IAD, he wanted to divest it of investigations and other related activities. He wanted IA to be done on a continuing basis. He did not want IA to be given additional responsibilities, excepting for some small jobs now and then in situations of compulsion.

The IAD in the MSEB had a strength of 103 persons. The expenses incurred on IAD worked out to Rs. 60 lakh. The function was headed by the director of internal audit, generally a retired IAS officer. He was assisted by a chief accounts officer holding an M.Com degree. One joint chief accounts officer and two deputy chief accounts officers were the other functionaries in the hierarchy. One of the deputy chief accounts officers was a chartered accountant. The chief executive agreed that an effective system of IA is an important instrument of financial control. However, due to certain limitations, it could not function as a powerful arm of the management in appraising the activities of the organization.

Though he wanted to associate the chief of IA as a part of his management team, he felt that the IAD staff was not creative in terms of thinking and lacked the necessary technical, communication and conceptual skills. He viewed IA as an independent appraisal activity and wanted to give the IAD a certain amount of independence. He wanted professionally qualified executives to work in the department. Preferring IA and AS officers he wanted the IAD to report to the managing director and did not favour surprise checks. He did not want any additional responsibilities, besides the IA tasks, to be given to the IAD. He was also against granting of any vigilance powers to the Department.

The director of IA revealed that no strict division was made between officers and staff employed in the IAD in terms of the functions performed by them, namely, financial audit, operational audit, systems audit, special investigations, etc. He agreed that the functions of his department should include review of the internal controls, examining the organization's records as to the implementation of top management policies, acting as a watchdog to prevent likely frauds, planning and carrying out checking of transactions, and verification of calculations. He disclosed that all the activities were combined with the accounts department, and that the status of the chief of IAD was equivalent to the head of any other department. He had a reporting relationship with the chief of the Finance Department. He was fully satisfied with the status, job content, and the power to perform his work. He said that he was conducting audit on a continuous basis throughout the year. The IAD did not contain any technical personnel. He disclosed that his department was conducting IA in respect of the various units once in two years, and that they were proceeding by test checking. Regarding the type of techniques used by IA staff while conducting the audit, he disclosed that these included confirmation, recomputation, examination of documents, enquiry, examination of subsidiary records, and correlation with related information. He expressed his satisfaction with the cooperation he received from his counterparts in the other departments.

There was divergence of opinion among the different operating managers regarding the role and importance of the IAD. However, taken as a whole, the operating manager did not want it to be a procedural consultant, detective agent, or an unwanted interrupter. Most of them felt that the management was underutilizing the talents of the IAD. However, they suggested that IA in order to enhance its capabilities, should staff itself with technical people. They were against granting of vigilance powers to the IAD. The operating managers agreed that IA reports should contain the comments of the respective heads of departments against the remarks made by IA before the report is submitted to the management. They wanted one representative from each department on the IA team. Most of them felt that the management was not fully utilizing the competence of the IAD. Some of them suggested that financial audit and systems audit should go together for attaining the ultimate aim. They wanted weightage to be given to practical aspects. Very importantly, unlike the other boards, the operating managers suggested that purchase, transmission and distribution losses, and revenue audit in respect of their functions could be taken up.

4.4.4 MPEB

While holding the IA to be an important instrument of financial control in the Board, the MPEB felt that it could serve as a powerful arm of the management in appraising activities of the organization only if the IA is concurrent and immediate. They wanted the chief of IA to be a part of the management team, and for the IA staff to be trained in management functions and for developing technical skills. They felt that the deficiency in professional qualifications could be made up through this training. Aptitude for the job was considered very important. They wanted the chief of IA to report to the board/audit committee. They also expressed the need for surprise checks to be carried out with regard to cash, verification of assets, etc.

4.4.5 NEEPC

The operating managers outlined the role of the IAD as that of an expert reviewer of operations, a constructive critic, a procedural reviewer, and a procedural consultant. However, they did not approve of its role as a detective agent. Whereas they expressed their agreement of undertaking IA of operating departments, they wanted technically competent people, having practical work experience and possessing sound knowledge of enterprise operations, to staff the IAD. They wanted transparency of the IA report which, in their opinion, should contain the views of the operating managers against the remarks made by the IAD. They were all for granting of certain vigilance powers and independence to chief of the IAD. They felt that there should be one representative from each department of the organization on the audit team. Unlike the other SEBs, they felt that the IAD helped enhance the effectiveness of the control systems in the organization and, therefore, they welcomed the role of IA in their operations.

4.4.6 RSEB

The operating managers suggested an intensive IA. In their view, IA should be more of a technical audit dealing with release of connections, checking of connections, position of transformers, checking of consumption variation register, checking of site checking register, position of physical compliance, checking of tripping of supply with log book, checking of load shedding register, checking of connected load register, nature of application of tariff, and theft cases. The function even included the checking of stores in regard to genuineness of requirement of material and its utilization, and availability of stock. On the revenue side, they suggested paying more attention towards average charges, wrong application of tariffs, wrong assessment of minimum charges, miscellaneous charges, less assessment of consumption charges, non-issue of first bills, and non-compliance of meter change and disconnection orders.

4.4.7 UPSEB

UPSEB had a full-fledged IAD, which was set up in 1977. The head of the IAD was a chartered accountant and a cost accountant. He was designated as controller of audit. He was assisted by a deputy internal auditor, who was also a chartered accountant, a cost accountant and an IAAS. These two functionaries were assisted by a chief accounts officer, deputy chief accounts officers (14), and accounts officers (16), accountants (18), auditors (79) and office staff (63). The chief of the IAD accepted that his department was conducting only traditional audit and his staff had no idea of systems audit. He also accepted that his department was not getting full-fledged support from the purchase and stores departments.

In the UPSEB, the controller of audit was of the view that the IA was an important instrument of financial control in the organization. It served as a powerful arm to the management in appraising the activities of the organization. He pointed out that IA's mission was confined to not only review of controls. He favoured the association of the chief of the IAD as part of the management team. However, he opined that the IA staff was not fully capable of thinking in management terms as they did not have the necessary technical, conceptual and communication skills. Despite this, he considered IA as an independent appraisal activity within the enterprise for the efficient functioning of the same. He opined that the staff in the IAD should be qualified, holding professional qualifications, such as CA, ICWA, MA (Economics), M.Com, Law, Engineering, etc. On being asked as to what type of qualification the chief of the IAD should possess, he indicated that those qualifying in the IAAS would be the best choice.

He agreed with the view that the IAD should comprise an interdisciplinary team. Regarding the authority to whom the department should report, the answer was that the Board Audit Committee was the best choice, though in UPSEB this function had to report to the director, Finance. He was all for exercising surprise checks, but was critical of the permission which had to be taken from line executives to conduct inspections of the operating units. He observed that sometimes the functions of the IAD could extend to investigations and wage board negotiations. He recommended that IA should be conducted on a continuous basis. He did not want the IAD to be burdened with any additional responsibilities arising in the shape of budget compilation and its review, liaison with statutory auditors and management, assisting with the compilation of annual accounts or helping in clearing out the backlog of accounts. He did not want any vigilance powers to be given to the IAD. The controller of audit revealed that the existing strength of the department of 192 personnel was adequate.

However, he indicated that the IAD did not have qualified internal auditors. Regarding maintenance of rapport with other departments, he felt that it was good only to some extent. On being asked about the functions of the IAD, he revealed that they comprised reviewing the internal control systems, verifying the financial accounts and vouchers, and physical verification of assets. Interestingly, appraisal of the activities of UPSEB in terms of policies and procedures was not considered as a function. The composition of the IAD disclosed that only accountants were part of the staff. During the discussion, it was further pointed out that regular IAs helped in checking frauds and misappropriation and in exercising effective internal control. However, it did not help in enhancing the effectiveness of managerial decisions. The cases of defalcations and defrauds brought to the notice of management by the IAD were on the increase. In UPSEB the expenses incurred by IA were about Rs. 175 lakh per annum. The type of punishments imposed for frauds and misappropriation included stoppage of increments, stoppage of promotions, and imprisonment. The rights enjoyed by the IAD included inspection of accounts and vouchers, making surprise checks without prior permission, making a report on any individual in the case of negligence and misappropriation, and recommendation of action against such individuals.

The operating managers wanted IA to be a constructive critic and acknowledged the need for its presence. They were, however, not in favour of assigning vigilance powers to the IAD. They wanted the department to be given adequate independence within the administrative set-up of the undertaking. The reporting relationship was to be with the managing director. They suggested the hiring of qualified technical people on the IA team, with sound knowledge about operations, and at least one representative from each Department of the organization to be in the team. They felt that the IAD was of little help in enhancing the effectiveness of the managerial capability in the organization.

Further, they hinted at the need to improve the technical competence of the IAD staff. They wanted the staff to be exposed to the practical working of operations. While that IA was felt to be essential for an organization to evaluate the efficiency of the system, they also wanted the comments of the respective heads of departments to be evaluated against the remarks made by the IAD before the report was submitted to the management. Whereas undertaking operational audit was considered the responsibility of the IAD, they wanted only competent people to staff the IA function. However, as this condition could not be met, they were skeptical about the success of any operational audit by IA.

4.5 CONCLUSION

IA is an important component of OPTIMA. This chapter shows that this function is in very bad shape in the SEBs/unbundled power utilities. The concept of IA has not taken deeper root. IA has not been organized suitably in terms of its location in the organizational charts and its functional relationship with other activities. The IA function has also been inadequately staffed. Professionals have not found their way into IA set-up. The existing incumbents, on the other hand, have not been exposed to any training. The budget allocations to IA as a percentage of total turnover have been very insufficient. IA charters lack clarity and the role of IADs has been relegated mostly to internal checks. The concept of IA being systems audit has not found much favour. Audit committees have not been set up by the power utilities and their role as a mechanism to discharge corporate governance effectively has not been much appreciated. Though the presence of a strategic audit plan is a primary requirement for the success of IA function, IA organizations have not appreciated this significance. The IA coverage has been very narrow and not exceeding 50-60 per cent of the total transactions. The reporting arrangements lack harmony. The IA function mostly reports to controller of accounts or officials attached to director of finance. The use of contract internal auditors has not made inroads. As IA is almost invariably inadequately staffed the number of IA units also fall short of organizational requirements. IA staff usually does not interact with the operating staff which naturally have a divergence of opinion about the role and importance of IA. The technical competence of IA staff is completely unsatisfactory

REFERENCES

Mishra, R.K., *Internal Audit in State Electricity Boards With Reference to Andhra Pradesh State Electricity Board*, Hyderabad: Institute of Public Enterprise, 1999.

Mishra, R.K. and B. Navin, *State Level Public Enterprises in India*, New Delhi: Mittal Publications, 2001.

Mishra, R.K., P. Geeta and B. Navin, eds., *Some Issues Relating to Public Enterprises Management*, New Delhi: Anmol Publications, 2001.

The *New Indian Express*, 4 Sept. 2001, Hyderabad edn., p. 11.

5

Profit Centres as Instruments of Performance Enhancement of SEBs

V. RANGANATHAN

The Electricity reforms in India started at the generation end with private sector participation in the form of Independent Power Plants (IPPs) taking investment and selling power to the 'single buyer' SEB through a Power Purchase Agreement (PPA). This was accompanied by unbundling of generation, transmission and distribution, corporatization of the unbundled entities and setting up of regulatory commissions at the central and state levels. However this model had a fatal flaw, viz., the 'single buyer' SEB was financially bankrupt. This resulted in a high payment risk for the IPPs which, along with the cost plus guaranteed return enshrined in the power ministry's notification resulted in high cost PPAs. The State Electricity Regulatory Commissions (SERCs) became toothless[1] against the Government owned SEBs or their unbundled and corporatized avatars, unable to rein in the huge transmission and distribution losses, a big part of which was plain theft in collusion with the SEB staff . With the advent of SERC's the SEBs now manipulate T&D losses in the *opposite* direction. Earlier the SEBs understated the T&D loss, booking a major part of the commercial losses under agriculture consumption since agriculture was not metered. Now they claim artificially high figures in order to get tariff increases.[2] After claiming 21 per cent T&D loss for several previous years, the Karnataka Power Transmission Corporation (KPTCL) claimed 38 per cent T&D loss for the year 1999-2000, in its first tariff filing before the regulator KERC. Several other SEB's have exhibited similar behaviour.

The second problem is the tariff subsidy to agriculture, which though overstated in the SEBs' books till recently, nevertheless erodes their financial viability considerably. Several studies have documented the fact that the increased incomes of farmers from electric lift irrigation do enable them to pay the long run marginal cost of electricity supply.[3] Still, political leaders find it expedient to continue the subsidy, and expect the SEBs to get the resources for this from within themselves, through cross subsidy from the industry. At present the electricity sector is left with:

— IPPs with high cost PPAs which prevent entry of low cost IPPs
— A financially bankrupt single buyer SEB or its corporatized version
— A regulatory set-up which—in the absence of privatization—has nothing to regulate, but only adds to the cost
— An unreformed SEB which has gone through the motions of 'reform' after endless studies by multinational consultants
— A near total drying up of funds from the central government in the expectation of the privatization prince to arrive.

The only relieving factor is that the industry is in recession so that the supply-demand deficit has not reached alarming proportions, but this relief may be temporary.

There are two clear reasons for the tardiness of reforms. The first is starting at the easier end, viz., generation leaving the intractable issues of distribution to a later date. The second is the World Bank's emphasis on privatization without paying sufficient attention to competition, which has resulted in high cost PPAs and Enron-like issues slowing the pace of reforms. In this chapter we take up the reform issue in distribution and address the scope for performance improvement under public ownership primarily through the introduction of the profit centre concept.

5.1 SCOPE OF A PROFIT-CENTRE CONCEPT

Maharashtra State Electricity Board, possibly for the first time in the history of SEBs, started getting figures for the distribution losses for its urban divisions. This was under the presumption that the arbitrary booking of unmetered agriculture consumption would be minimum if not nil in the case of the urban feeders and the resultant distribution loss would be a true reflection of the technical and non-technical losses, i.e. free from the errors causes by assessment of unmetered consumers. The findings were revealing. The Kohlapur

TABLE 5.1: MSEB DISTRIBUTION LOSS PERCENTAGE IN URBAN AREAS 1998-9

Urban area (Division)	Loss %
Kohlapur	6.98
Ichalkaranji	9.30
Sangli	9.67
Pune Urban Zone	13.31
Nasik	17.27
Solapur	18.08
Bhandup Urban zone	20.24
Akola	21.27
Amaravathi	22.41
Malegaon	23.22
Jalagaon	23.69
Aurangabad & Beed	26.43
Dhule	28.60
Nagpur	30.44

division reported the incredibly low distribution loss of around 7 per cent, a significantly lower figure than privatized and denser areas with more concentrated loads. Table 5.1 gives the details.

Since the technical characteristics of the distribution system in these divisions could not explain such a wide variation, the only explaining variable is the amount of non-technical losses. The scope of introducing the profit centre concept lies here, as this exercise is able to isolate the high loss divisions from low loss divisions. The second step would be to study the low loss divisions and discover the secret of their success and try and see if the success virus can be transplanted in the rest of the system.

In the Kohlapur case our study revealed certain aspects of control systems that were differently done from the rest of Maharashtra. The cadre rivalry between the engineering and accounts branches in SEBs is well known. No member of the engineering branch would work under the supervision of the accounts branch. But a task like disconnection of consumers who have defaulted payment involves close coordination between the accounts staff who first detect the default, and the disconnection crew who have to perform the disconnection. In Kohlapur, the disconnection crew was willing to work under the accounts supervisors and this lead to timely and effective disconnection. This in turn lead to better revenue realization.

TABLE 5.2: MSEB PERFORMANCE ZONE-WISE 1998-9

Zone			Collection	% meter not read		Performance	
	% Un-metered	Loss %	%Target unmet	Resi-dential	Industry	Wt.Av. 3-6	Rank
1	2	3	4	5	6	7	8
Rural							
Amaravathi	35.0	20.74	12.88	17.45	13.89	36.8	3
Aurangabad & Beed	61.0	24.64	23.44	22.14	22.01	49.3	6
Kalyan & Konkan	15.4	23.50	6.84	28.3	21.68	44.9	5
Kohlapur	38.0	14.23	6.29	10.23	21.45	29.9	1
Nagpur	5.5	15.07	16.98	20.56	20.58	36.6	2
Nasik	58.0	22.13	14.49	28.75	17.41	43.8	4
Total Rural	37.3	20.26	11.85				
Urban							
Bhandup	0	17.28	9.5	22.72	22.50	38.2	3
Pune	0	10.08	11.36	17.13	29.44	33.3	1
Nagpur	0	21.10	0.27	20.31	19.62	37.1	2
Total Urban	0	15.33	6.48				
Total MSEB		19.43	10.47	20.80	20.82		
UPSEB		24.41	17.00	47.11	22.03		

However, the hypothesis that higher unmetered consumption will encourage the tendency to show industrial theft as agriculture consumption and thus show reduced losses is not established. For, there appears to be no correlation between assessed consumption as a percentage of total consumption in each rural zone and the loss per cent of the zone (see Table 5.2; columns 2 and 3). Also the urban zones account for two-thirds of the consumption but have a lower loss than the rural divisions, which account for only one-third of the consumption and account for a higher loss.

Among the rural zones, Kohlapur again emerges with a high agricultural load (column 2) and low losses while the Kalyan-Konkan Zone emerges as the one with a low agricultural load and high losses. Nagpur Zone has a low agricultural load with low losses while the Aurangabad and Beed Zone has both a high agricultural load and high losses. Among the urban zones, Pune emerges best followed by Bhandup and Nagpur.

Two more performance measures show up, viz., collection rate and meters not read among domestic and industry consumers and the cumulative performance is seen as a weighted average scores of columns 3 to 6. The zones are ranked in the last column based on these weighted average scores. Present weights are however judgmental and can be changed; still they give a clear discrimination between the zones.

The distribution loss figures for UPSEB are given alongside. Tables 5.3A and 5.3B give a comparison between MSEB and UPSEB in terms of the frequency distribution of the zones and areas with varying losses:

Table 5.3A shows that while in the MSEB the poorly performing zones are as many as the good performing zones, in UPSEB the very poorly performing zones are somewhat in a minority compared to the poorly performing zones. However, in urban areas, the inefficient areas are in a majority in UPSEB. Thus the task of cleaning up in the urban areas is more daunting in the UPSEB and requires an iron will.

Recently, the Karnataka Power Transmission Corporation played a different type of game. It claimed a *transmission* loss of 23.6 per cent for 1999-2000, before the Regulatory Commission, against a norm of 6-8 per cent.[4] There was no way of verifying this figure, because in several places there was no meter even at the feeder level. An independent study initiated by the KERC estimated a significantly lower level of loss around 16 per cent by combining the meter reading data with load flow study. By overestimating the transmission loss the KPTCL could be giving an impression that

TABLE 5.3A: DISTRIBUTION LOSSES 1998-9
NO. OF ZONES IN EACH LOSS CATEGORY

	Distribution loss %			
	<16	16-21	>21, <=35	>35
MSEB—rural	2	1	3	
MSEB—urban	1	2		
UPSEB—rural		7	4	1
UPSEB—urban		1	1	1

TABLE 5.3B: DISTRIBUTION LOSSES URBAN AREAS: 1998-9
NO. OF AREAS IN EACH CATEGORY

	Distribution loss %						
	<10	10-20	20-4	25-9	30-4	35-9	40-9
MSEB	3	3	5	2	1		
UPSEB			1	2	7		4

this is an 'uncontrollable loss' implying that theft, which is usually associated at the distribution level, is less. Requiring the Board to have compulsory meters for all feeders both at incoming and outgoing levels for each division could eliminate such manipulations by the Board, as well as identify the culprits who are responsible for the revenue leakage.

5.2 IMPLEMENTING PROFIT CENTRES

The Board could view each division as a profit centre. Each division could be asked to provide data as per Table 5.4 to the head office.

Columns 8 through 11 give the performance of each division. By circulating the division-wise performance to all divisions (through the internet) every division chief is assessed by all other division chiefs and this creates what in audit parlance is called audit anticipation effect. Thus the first incentive for better performance is to stand up and maintain self-respect in front of the comity of colleagues. In the case of Maharashtra and UP even though the variation in performance was available to the management, there appears to be no system to give feedback of this performance of everyone to everyone else.

The second step is to arrive at a concept of **surplus** for each division. Surplus is arrived at by adding the cost of received electricity and the out of pocket cash outflow for the division for the month and subtracting it from the cash inflow. The cash inflows are proceeds through sale of electricity, and the cost of received electricity

TABLE 5.4: DIVISIONAL OPERATING STATEMENT

	Energy in MU			Revenue in Rs. lakh						Revenue Arrears
Divn. no.	Recd.	Metered	Sold	Billable	Billed	Realized	Deficit in billing (5)-(6)	Current arrears (6)-(7)	Distribution Loss	
1	2	3	4	5	6	7	8	9	10 [(2)-(4)]/(2) %	11 Equivalent # months revenue

is the product of the number of units of electricity received by the division and the transfer price of electricity received at the pooled cost of generation and purchase, and an adjustment factor for the load mix. This can be called the load mix factor. The load mix factor will compensate for lower tariff loads by reducing the input cost of power for the division. To arrive at the surplus, division-wise energy accounting is a necessary condition. Now this surplus figure too could be circulated to all divisions. Divisions that make a surplus beyond a stipulated amount could be given resource discretion. That is, they could be permitted to retain the surplus for their budgeted and authorized expenditure, instead of repatriating the entire revenue to the head office and wait for the expenditure amount to come from the head office. This would motivate the division chiefs to improve their load mix, by giving new connections faster, etc., so as to further increase the surplus.

The third and final step is to establish each division as a profit centre, clearly providing them the depreciation figures of their assets and the overhead allotted. Once the profit for each division is calculable, then a financial incentive system could be developed so that every one in the division is rewarded for their efforts towards creating the profit, in proportion to their efforts.

5.3 CONCLUSION

Having established such a system, three questions have to be answered:

1. What type of incentives should be offered to the 'overheads', viz., head office and common service personnel?
2. What should be the disincentives for non-performing divisions?
3. What is the guarantee that the incentive system will work if the existing system promises more rewards, as it undoubtedly does?

Before answering the first question, the common malady of overstaffing at the head office should be rectified. To that extent, greater incentives for field people would be a step in the right direction. After that, however, SEB's performance-based incentives could be extended to the head office staff. Incentives for common services could be worked out in their transfer price. At the same time, the operating divisions could be allowed freedom in deploying their services based on their own cost benefit of the services offered, on a competitive basis. In other words, divisions may use internal services based on an internal competition.

Non-performing divisions could be handled by giving them a long rope, by management resorting to judicious transfers, isolating the effects of bad employees by making them redundant, etc.

As regards the third question, the entire incentive-based profit-centred approach is designed upon the premise that marginal utility of earned incentives would be more than marginal utility of ill-gotten income for most employees.

The above scheme was suggested to the Nepal Electricity Authority as part of its commercialization. The board even decided to adopt it after extensive discussion. However, it is not known whether it was actually ever adopted. The time may have come for the SEBs in India to try it, as the only other alternative would be privatization.

NOTES

1. V. Ranganathan, 'Proliferation of regulatory commissions: Putting the cart before the horse', The *Hindu Business Line*, 15 April 1999.
2. V. Ranganathan, 'Metering as a way to reform SEBs', *Economic Times*, 17 September 2001.
3. See for instance, V. Ranganathan and T.V. Ramanayya, 'Long term impact of Rural Electrification: A study in UP and MP', *Economic and Political Weekly*, vol. XXXIII, no. 50, 12 December 1998.
4. KPTCL's tariff filing before KERC, dated 20 September 2000.

6

Delegation of Power in SEBs under Management Contracts

JOËL RUET

This chapter deals with the structural impacts of the internal organization of the State Electricity Boards (SEBs) on their efficiency, and proposes some solutions to modify the situation. It is based on fieldwork and in-depth interviews. I interviewed nearly two hundred people belonging to SEBs, ministries, governmental agencies, regulatory commissions, private groups, multilateral agencies, etc., ranging from meter readers to power secretaries, in seven states—Andhra Pradesh, Delhi, Haryana, Kerala, Orissa, Punjab, Rajasthan.

The chapter focuses on two issues:

(i) in the budgetary and reporting procedures, what are the factors that structurally alter the decision-making process and lead SEBs to inefficiency?
(ii) taking stock of previous experience, which proposals show a minimal credibility and sustainability in improving the organizational parameters of efficiency?

I argue that, because of the budgetary and information systems, decision-making cannot rely on sole cost-benefit analysis resident in them, and the SEBs are by nature alien to the question of productive efficiency. Consequently, what is required is for a radical transformation of the internal organization of SEBs, and the way decision-making is structured in them. The argument holds that efficiency-led decisions may actually be made under a scheme where financial power is decentralized. Such a tool also shows a higher credibility/ sustainability under a Management Contract scheme, with specific conditions that are detailed below.

6.1 STRUCTURAL ROOTS OF INEFFICIENCY

It is a known fact in India that SEBs are subjected to the discretionary power of the state. And beyond the widespread statement of 'day to day interference', the very internal organization of SEBs is affected by this fact. The state, by political nature, simultaneously pursues various objectives which are non-reconcilable through monetary parameters. These objectives can never be decentralized at the SEB level, and definitely not inside the SEB. In that context, the discretion of the state is not an 'interference' but is, on the other hand, structural. Of course, the discretionary power of the state is not used in a *systematic* way in every decision, but its possibility leads to a *systemic* uncertainty on the effects of decisions taken inside the SEB. Indeed, the possibility of an *ex post* discretionary decision leads to the impossibility of an *ex-ante* cost-benefit comparison for the decision-making. Thus, the latter can be more rational and more stable by following procedures, that is, procedural rationality. As far as reforms are concerned, one has to realize that, in organizational terms, this procedural rationality is sustained by a budgetary structure that denies any actual delegation of discretionary power to the hierarchy inside the SEBs. Indeed, the so-called budgets annually allocated to each unit are still dependent on the discretion of the headquarters as far as their actual use is concerned.

6.1.1 The Internal Organization of SEBs

Let us analyze the budget as a tool for allocating discretionary power. It is based on public accountancy rules. The expenditures involve three categories, (i) the 'capital' section, related to new constructions, (ii) the 'establishment' section, related to staff expenses and administration, and (iii) the 'revenue' section, related to maintenance. There is neither the possibility of diverting funds from one of the budgets to another (capital, revenue or establishment), nor to divert funds within a type of budget from one activity to another. This means that, in this system, unspent allocations will not be transferred to the next year. In addition, it is at most a partial equivalent between budget sections and not a general equivalent, since inter lines and inter types transfers are not available. Hence, once the budget is decided, the power of arbitrating between maintenance and new works is not in the hands of the field. These are classical facts in a public accountancy system.

But going beyond, the relevant point is that there is no direct allocation of actual cash to the field. Once the budget is decided, physical inputs of the field are provided directly from the Board, on a case to case basis, on duly argued requests by the field, following a process which begins at the moment the needs occur. In practical terms, at least three levels of administrative sanction are required for any expense. Next, the requirement is transmitted to the headquarters, who grant the expense. No cash is given, however. Instead, the inputs might be directly—physically—provided by the board or a different unit. So-called expenses are in fact cleared at the level of the headquarters in accounting books. Even where this is not the case, or where the market supplies the inputs, the board will still not directly allocate cash to the field. In this latter case, a procedure is followed in which the supplier is selected during the phase of clearing the administrative sanction, and the board notifies the local branch of the State Bank of India to give the money. The only difference in the two types of supplying is the interlocutor who is dealt with—with some implications regarding the type of the procedure—since any supply from the market requires a tender notice. The operation of the 'capital' budget (new works) is even simpler. Material is transferred to the stores, then to the works department that is building the new equipment, and subsequently asking the field for a certificate of acceptance. The store does not either 'buy' or 'purchase' the materials, as the board only notifies the bank to pay the supplier. In addition, no store—attached to a specific area—is the 'owner' of its stock, as any store from any other area might require some of its materials. As far as collections are concerned, they are directly transferred to the headquarters.

To sum up, whatever the action undertaken by the field, no actual money comes into the hands of the field officers.[1] Indeed, the so-called budget allocation does not even allow the discretionary power to arbitrate within two similar actions in slightly different places or at slightly different times. Any decision is separate from any other one, and each follows a separate path. This means that, beyond its general characteristic of public accountancy, the budget does not play the role of a general or even partial equivalent in exchanges. Indeed, even within the already constrained framework of public accountancy, the goal of managerial freedom to make partial decisions on actions and time is unmet. This is a book-currency budget, or a paper budget; it is not actually allocated.

Let us now look at that issue from the perspective of headquarters. In an enterprise, the budget is classically designed to be a managerial tool for the headquarters, allowing a monitoring of the totality of expenses, while decentralizing the budgetary constraint, and not bothering the headquarters with day to day decisions. This decentralization can be possible only if the credits in the budget serve the purpose of being—at least partially—monetary equivalents between the actions one needs to arbitrate. The paper budget at stake in SEBs simply precludes this kind of decentralization. More generally, virtually every decision in the SEB has to be taken collegially, between at least three levels of the hierarchy, often with the involvement of the headquarters. This involvement is not required when clearly defined procedures can apply. This is for instance the case for some urgency matters, for which the officers have a strictly limited discretionary power. Typically, in many SEBs, the amount is at most about a few thousand rupees for a subdivisional engineer, who still has to immediately report for his expenses. Here again, the field has no formal discretionary power over the decision, as it is required to implement the procedure.

6.1.2 A Digression: Why Public Accountancy is Adapted for Some Sectors and not for Others

The point is not to argue that public accountancy is not adapted for every sector. As far as regalian functions or public health, education, or defence, etc., are concerned, the outputs can hardly be valued as being non-commercial. In these cases, public accountancy is useful to monitor the level of expenses. Also, if an activity mostly involving investments like roads is concerned, public accountancy is useful, as the commercial side of the activity is nil (road use being free). Second, an extra load on a road leads in the short term only to traffic jams, and not to additional costs to the road authority. On the contrary, power overloading affects the state of the technical system itself. So, power system planning, if adressed with an economic approach, implies the computation of avoided costs. In this sense there is no economic calculation comparing investments and avoided costs as far as roads are concerned. The public accountancy system is then useful for roads, where this question is not stringent, but leads to major cost inefficiencies in power sector. As far as power is concerned, public accounting is useful for electrification, but not for network operation.

6.1.3 The Reporting

The reporting can be qualified as non-integrated, meaning that it does not show in front to each others inputs to outputs, or relate the energy delivered, to the energy billed or to the energy paid. It as well precludes the emergence of decentralized management tools. As far as the accountancy is considered, IMI states that from the division level upwards, the accounting is a unified commercial accounting, but that, from the subdivision level downwards, the system is filled with 'statements and statistical information only, instead of accounting entries'. An examination of accounting in SEBs thus involves issues of completeness as well as of consistency. A former power secretary once declared to me that the book accounts of SEBs are 'a terrible mess, with no link with reality'. Ultimately, the whole reporting system is not designed to provide information to be used in decisions, but to offer an administrative reporting, allowing administrative inquiries. It can be qualified as non-integrated, since technical (repairs), physical (energy) and commercial (billing, revenue) aspects are neither put in front of each other in any report, nor in their respective accounting.

6.1.4 How Paper Budget and Non-integrated Reporting Result in Inefficiency: The Example of Distribution Transformers

At this point I would like to illustrate the structural effects of the internal organization of SEBs. Here is what happens when a transformer is damaged in a subdivision.[2] The SDO, with the clearance of the Xen, issues a store requisition. The store replaces it with a new (or repaired) one; following this, the damaged one is repaired in the workshop, before being returned to the store if it can be repaired, and then being ultimately returned to whatever subdivision that faces the damage.

Due to the absence of integrated information, the headquarters has no way of exercising control through techno-economic calculation. To monitor this activity, therefore, there is an administrative circuit, based on inquiries starting with the placement of responsibility for the burning, and checked by the hierarchy on a sampling basis. The SDO is required to inspect 100 per cent of the damaged transformers; the Xen 25 per cent of the 200 kVA and above; and the SE 50 per cent of the 500 kVA and above. Here things begin to get interesting. The headquarters can only monitor the situation through a constraint on the rate of damage. As potential administrative sanctions are in place, proving innocence becomes a big deal for field employees and managers. It is common practice to declare, in damaged transformer reports, that all the necessary protections were installed and that the operating conditions were normal. Hence the damage is attributed to an 'internal fault', and the report is signed by the SDO.[3] The reports, stating that the necessary controls have been made by the SDO, the Xen, and the SE, are further written up and sent, without any correlation with actual checkings—they are either not done, or are delegated to subaltern officers. Because of the scarcity of means, the field agents feel that they are not being able to control the rate of damage. Under normal operation, they have to fulfil the continuity of supply. The absence of meters deprives the agents of the tools they need to foresee the burning of transformers. Worse, after the damage has taken place, the most important task is to issue the requisition forms, and formally declare that the checkings have been completed, so that the paper work is done. The logic is not to foresee events, but to justify ex post that they had to occur and that the accurate procedures have been followed. This further allows bargaining on the non-realization of given technical targets. The reports will state that replacement of transformers is necessary due to day to day operations.

Hence the notion of cost never occurs in the decision. The transformers must be made to run for as long as possible. When they are burnt, and only when they are burnt, are they to be repaired. Finally, there is neither actual budgetary monitoring—based on costs—nor actual administrative efficient monitoring tools for the headquarters.

6.1.5 Making Cost-benefit Comparison Central in Decision-making

The budget does not allow any actual power on decisions, which are to take place through procedures of administrative sanction. Regarding the place in the hierarchy, there is no discretionary attribution of tasks, since every decision must go through a formal agreement at several levels. Managerial pressure is not available, as there is no strict allocation of power over human resources. Reports deal with technical matters, and their number is plethoric, but their information is useless in decision-making, since there is neither any worthwhile information-gathering nor control. Finally, there is no reliable accounting system; accounting rules are transgressed in many ways. The reporting system basically comprises ex post records of results and decisions for administrative reporting, i.e. for providing a basis to check the implementation of centrally-defined procedures. In this case, using the term 'information system' is simply not valid, as the term refers to the utilization of the information generated as an input in decisions.

Field inquiries show that, even after structural reforms (corporatization, de-integration, setting up of regulatory commissions), the decision-making procedures remain unchanged; internal discretion is still unmet, there are no rules for subsidies allocation, while executive instructions remain direct. Hence, organizations issued from SEBs are still not efficiency-led in their behaviour. This is not an issue of time, but, as shown before, the result of unaddressed structural reasons.

To sum up, the primary radical change that needs to take place is the institution of cost as a central concept and criteria for behaviour, as in a genuine enterprise. This will be the *sine qua non* condition in the pursuit of efficiency. To meet this aim, a true budgetary allocation has to be put in place, and the evaluation of agents has to be based on cost-related results. Subsequent training has to be based on the possibility to use this budgetary allocation, in order for costs to be integrated in behaviour. It is only after this actual allocation (that is after this step towards creation of cash flow rights) that the question of incentives and efficiency can be addressed. Indeed, previous analysis has shown that, as long as executive instructions remain, and as long as the discretionary power is not properly allocated, the system will not be radically altered. Put in other words, the measure of allocating budgets remains insufficient to the extent that executive instructions can expropriate the benefits of the decisions that are taken. After budget allocation, property rights are still not properly established. So, complementary to the establishment of a budgetary allocation, and in support of making central the concept and the use of cost, two necessary conditions arise.

First, discretionary power has to be established; this would be internal to the SEBs, and given to the Field officers. This partly means that the budget allocation or at least a part of it has to become available for use which is locally defined and decided upon.

That way, the cost vs. benefit analysis may diffuse into the structure. Conversely, setting discretionary power means putting in place an allocation of decision-making powers, restricted to few levels of hierarchy, and ideally for most of the common decisions, to a one-level structure. In this scenario, information is an advantage, but more as an accompanying reform than as a structuring necessary condition. Simple improvements are already available without any need of further information at the field level, regarding transformers, for instance. Also, once the focus has been shifted to costs, then it becomes rational for the field to put in place new procedures on information that will be efficiency driven. Section 6.2 deals with a related proposal.

Second, executive instructions from the state government have to be credibly removed, and replaced by general objectives which allow the SEB freedom to achieve them in an accountable manner. Of course, this issue raises many questions about the feasibility of moving from a regime of executive instructions to a well-policed one. Section 6.3 proposes a management contract scheme with clear and stringent conditionalities for that to be achieved.

6.2 GOING BEYOND PROFIT CENTRES: FINANCIAL DELEGATION OF POWER

This section makes a proposal regarding the establishment and decentralization of formal discretionary power within the SEB, especially regarding financial responsibility. Putting this in place, it is suggested here, is a necessary condition for breaking away from the classical system. This section then focuses on a key issue unfortunately being neglected in current reforms. Let us again take the example of transformers operation to be easily generalized to the whole of operation and planning. Let us then show how this proposal goes beyond the sole establishment of profit centres.

6.2.1 Rationale and Method

Since the present system of centralization of the budgetary demands does not help to provide a tool for actual management and control to the headquarters, structural reform should look at decentralizing a part of it to the field. The circle level seems to be the most suited as the first step for such a reform. This would allow the field to acquire property rights based on the results it achieves. Each circle would be more accountable as long as a techno-economic approach becomes available. Put in place jointly with a system that fixes actual upper limits of expenditures and lower boundaries for revenue collection, it would constitute for the headquarters a guarantee for the respect of financial constraints, which in interviews appears to be a limiting factor on the will to implement decentralizing measures. Of course, pure zonal profit centres are hardly expected to be a viable solution (because of an uneven mix of consumers, and the present global financial situation). Hence, this decentralization would be limited, while being progressive enough to accompany the necessary learning process.

Let us look at a loss-making zone (refer to Figure 6.1) for the proposed mechanism. The budget of year N+1 is derived from actual expenditure of the previous year (for all divisions in a given circle), plus an applicable increase to establish the new budget. From that level, let the headquarters allow to the circle a fixed part in cash.There are two ways of doing this. One way is through 'direct allocation'; headquarters should start giving cash allocation to the field, regardless of initial improvements in revenue collection. The other way is through 'bound allocation', by waiting for initial revenue recoveries. In Figure 6.1, this cash portion is represented in the two parts of the budget. A discussion of their respective pros and cons follows below.

The cash allocation has to be done according to a given timetable. To start the reform, it could be concerning the lines of credit

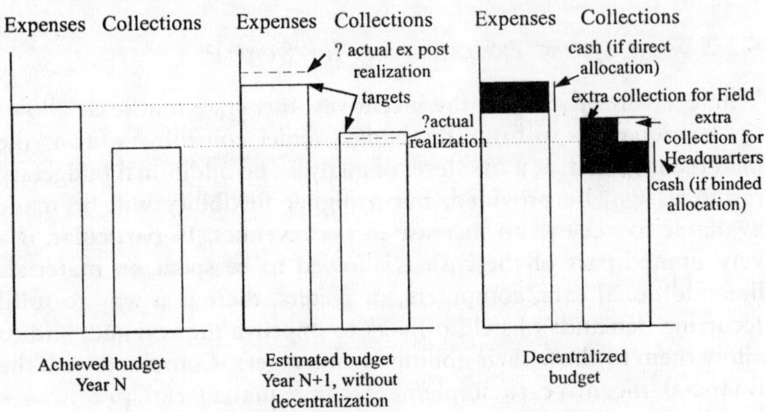

FIGURE 6.1: A PROPOSAL FOR BUDGETARY ALLOCATION CUM DECENTRALIZATION IN SEBs

corresponding to the investments, maintenance, and repair of transformers, as they constitute a crucial point. The headquarters could authorize the field to use:

— either this cash in these transformer-related fields, irrespective of the share between them, i.e. to divert from previous use within these budget lines
— or to use a limited part (up to a fixed limit, say a few per cent) of this cash for any input purchasing.

This creates property rights for the field: it will be able to use the cash as it wants, when it wants. For the headquarters to be sure to control the total budget, it may fix the maximum limit on it, i.e. for the paper-budget part. This, in turn, solves the concerns of the headquarters related to the budget constraint of the field (on Figure 6.1, the level of expenses is fixed and cannot be higher). However, this can be balanced by personal incentives on wages in compliance with these limits. As far as revenue is concerned, let the headquarters specify a minimal revenue collection, based on feasible targets, for instance previous achievements. To be credible, the headquarters can make their cash disbursement subject to completion of a given timetable in collection. Above this level, let the headquarters allow the field to retain, as an incentive, a part of the extra realization for its own cash account (this measure can be additional to a personal incentive scheme), say on a fifty-fifty basis. These changes once made, incentive schemes have a high chance of becoming effective.

6.2.2 What is to be Expected from this System?

Whatever the efficacy of the incentives, this system at least allows the headquarters to bring the budget under control. As far as the field is concerned, at a first level of analysis, no additional budgetary resources will be provided, but a higher flexibility will be made available to achieve an increase in the revenues. In particular, if a very limited part of the cash is allowed to be spent on materials like additional cars, computers, or meters, there is a way to fulfil recurring demands of field officers to improve the revenue, and to allow them to show their abilities as managers. Combined with the financial incentive to implement some managerial practices—experiments which have proved to be efficient (rotating meter reader teams, for instance)—this should allow them to improve collections, and hence cope better with the associated budget constraint.

At a second level of analysis, the partial decentralization of cash at the circle level (unlike at present, when the budgets are specified divisionwise) provides some additional flexibility. This mutualization of the cash portion—or this local centralization—results in temporarily affecting the cash in some zones. This will allow better use of the cash where the investments are the most needed (the field being expected to have more information regarding this data than the headquarters). It can also be argued that, given the present high rate of damage in some states, or more generally speaking, the low level of collections, completing the enhancement of some specific zones is better than dispersing the efforts.[4] In the case of transformer repair, the resulting higher income (manifested by a lower number of transformers to be repaired in these zones, hence a higher number of transformers that can be replaced in other zones) could be used in other zones.

In the present system, field officers have to focus on a sole technical parameter—the rate of damage—trying not to exceed a given limit, first with no reason to try to achieve better, and second with paper work behaviour on that target. This reform would allow them to calculate that maybe, by admitting an initial disequilibrium between different zones (which they are not allowed now), they could ultimately reach a better result with the same cost. This would allow the field managers to see that they can implement the technical solutions they know, without any additional burden on the budget, or in the present system, any additional physical means in the long term. An additional practical reason for decentralizing at the circle level, is that experience, seniority, accounting and computing resources are higher at the superintending engineer level. These parameters would be important not only in managing the cash but in ensuring that the budget constraint is respected. As far as the headquarters are concerned, this reform would help them decentralize the burden of the financial constraint.

6.2.3 Practical Issues to be Solved

— Setting up a system of actual cash implies the implementation of a budget constraint. This has to happen in the framework elaborated below:

(i) A transparent and synthesized information system is in place, allowing both managerial tools for the field and control tools for the headquarters.

(ii) A coordinating committee between the field and headquarters is established, to follow the monitoring of the budget constraint.
(iii) Training of the SE and the Xens is organized, with accounting support from the headquarters.

—In particular, corruption problems might occur. Under this system, it would not be easy to divert the money directly to personal accounts, as there will be some control at least. However, a proper mechanism needs to be instituted, to ensure that managers do not enjoy siphoned funds physically, by buying unnecessary inputs. This percentage allowed for inputs purchasing might either be introduced after a while, or could be brought under the control of the monitoring committee through specific reports on their final profitability.

—*A priori*, direct allocation shows a higher commitment of the headquarters to the field, but the bound one offers an incentive to make collections faster. A mix of both, or initiating the reform with direct and then replacing it with bound allocation, could be the optimum option.

—The introduction of actual budgetary allocation must give disincentives. One needs to take into account the fact that unexpected events can occur, like a heavy monsoon, for instance. If that happens, the pilot zone shall of course benefit from comparable help to the other zones.

—The introduction of cash at the circle level also implies the introduction of cash in the store and workshops, with guarantees given on timetable and costs, and/or authorization to purchase from the private sector in case of non-compliance with the commitment (under the ex post control of the coordinating committee).

—As far as annual renegotiation is concerned, in order to maintain the incentives, some guarantees that the improvements achieved by the field will not be expropriated need to be given.

(i) The growth of the total budget shall be comparable to the growth of the total budget of the entire board.
(ii) The share of the cash part in this total budget shall remain at least constant, conditional to efficiency improvements.
(iii) The new minimal limit for revenue collection shall increase, but shall remain inferior to the actually reached level. This will ensure that the field gets to enjoy a part of the extra

revenue it managed to earn over a longer period of time, thereby strengthening the incentives.

These guarantees should be given for a period of at least three years, while this process is broadened to cover the entire area. The sole risk is that, over this period, the pilot zone achieves big improvements while the other zones remain in need of money. This risk is very limited. First, it relates to a positive context, where improvements have been achieved. Second, the number of profitable operations and improvements is enough to ensure that money coming from initial improvements will still be profitable in new ones. Third, the profitability in zones where actual improvements have not been initiated is only potential, and the incentive given to these zones by such an example is higher. Fourth, in this context, the headquarters receive more cash than in a business-as-usual scenario, which can be used for other areas (assuming that they will have actual property rights on it).

6.2.4 A Discussion on the Scope of Creation of Financial Discretionary Power

The possibility of starting such a scheme within a system that would remain the same in other respects has already been discussed, by pointing out the need for a specific monitoring committee. But another limit has to be studied. Indeed, the proposed scheme jointly addresses an expense issue—the burning of transformers, and a revenue one—collection improvement. The point up to which they are to be linked or separated has to be discussed. The stake is that, if they are not linked, then one would be tempted to look at easier schemes to allow collection improvement. Let us argue that they consitute a rather minimal package.

First, if the will is to generate and spread out a costs vs. benefits behaviour, then direct earnings tell more than indirect earnings through savings on burnings. Similarly, if the field accepts this scheme for reducing the transformer burnings and builds it into his behaviour, the access to additional sources of income may speed up the process. In return, there is an incentive on improving the collection. Allocating discretion and accountability on transformer repairs is profitably complemented by allocating discretion on collection. Improvement in collections is in this way simultaneously a catalyst and by-product of improvement in operation.

Also, allocation of discretion in collection cannot be started on an exclusive basis. Indeed, even if actual discretionary power is allocated in implementing improvement measures, there is no incentive in collecting more per se. If the additional income cannot be used in the field, then additional collection is useless. Similarly, the field being concerned with technical improvements, in the proposed scheme, it can consider the financial parameters an indirect way through which it is able to improve the main technical parameter. Whether there is a primacy of finance over technique or vice versa (which is rather a matter of type of 'culture of enterprise'), what matters is that decisions are addressed in a joint financial and technical approach, and no longer in a technical one only (the organization then acts as an enterprise on these matters, whatever its culture).

6.2.5 Decentralization Schemes Compared to Profit Centres

Some authors, for instance Ranganathan or IMI, support the decentralization of power in the power sector.[5] IMI advocates (i) 'commercial accounting', with an integrated and computerized system and with primary accounting set at operational units, as well as (ii) responsibility accounting, either in the form of cost centres or of profit centres. The actual difference between the two concepts is that in cost centres, only the direct costs are to be recovered, whereas in profit centres, direct costs plus depreciation costs (or costs of capital) are to be recovered. Ranganathan and Gupta are more precise on whether to use one or the other. They conceptually separate support activities from operating divisions. They state that a 'decentralised organisation structure with responsibility and accountability at lower levels', has to be promoted and detail that 'this is achieved by incorporating a responsibility centre system in the organisation in terms of cost centre for support activities and profit centre for operating divisions'. This provides an information system. But these accountancy issues are hardly a new measurement system, if not sustained by the necessary discretion to actually use it. Both Ranganathan and IMI base their proposals on a necessary delegation of authority. How does this proposal of decentralization fit with the need for allocating discretionary power? Mainly, there is an aim at creating control rights and allocating them to the field units. However, nothing is granted since the executive instructions of the concerned state are a priori not altered. Also, to be fully allocated, these rights must also be credible in terms of technical possibility for the field to make use of them. For instance, if the decentralization means that the field is granted higher amounts of so-called delegation of power under a remaining system of paper budget (i.e. with the need to apply for whatever expense), the progress is nil. This means that, as long as the paper work and the following of procedures remain, the setting up of these control rights is not credible. This is where the proposal of delegation of financial power actually comes in to supplement this profit centre concept.

Indeed, the issue of cash flow rights demands special attention. For instance Ranganathan, in a very similar context of Nepal, proposes a transition process, during the first phase of which, a unit becomes a cost centre, and then a profit centre in the second phase. And he states that only 'the branch managers who make surplus in phase one and profit in phase two will be rewarded with budgetary discretion'. In other words, property rights will not be set up straight away. However, it is difficult to imagine how surplus and then benefits are attainable in the absence of a radical change. It seems far-fetched to hold that these goals could be achieved with the setting up of a measurement system. And practical experiments prove deceptive in that respect.

For instance, profit centres have already been set up in Kerala. The key element of the Kerala State Power Policy, announced in August 1998, is the restructuring of the KSEB through profit centres. There are now three profit centres: generation, transmission and distribution, which itself has three distinct units covering the three geographical regions: north, central and south. This new structure has been set up further to the drafting of action plans suggested by a cross-section of over 120 KSEB employees from different categories. However, field inquiries conducted as late as in December 2000, showed that no actual change in the paper work has occurred. The concept of profit centres, in place of setting up managerial tools, has been turned into an accounting system, hardly measuring physical inputs and outputs, in a context of unset intermediate tariffs and bundled capital costs. As far as the external discretionary power of the State is concerned, it has hardly changed.

From the above discussion, it seems that profit centres should be targeted directly, and soon upgraded through actual delegation of financial power. Ranganathan rightly discusses the sustainability of such schemes. He recognizes that the proposed scheme 'makes an assumption, i.e. the top management, represented by the board, has only one objective, viz., achievement of business efficiency and that there is no political interference on the firm's top management in pursuing this goal'. The administrative nature of SEBs of course proves this to be hardly feasible.

In this context, let us discuss the role of the state, and advocate a specific type of management contract.

6.3 MANAGEMENT CONTRACTS AS INFORMATION-BUILDERS

Given the past experience in India, many observers raise doubts about the credibility or at least the sustainability of attemps to carry out internal reforms that are made by the various state governments, and advocate privatization as the only solution. Privatization in the context of administratively-led SEBs has already been discussed at length elsewhere (Ruet, 2001a and 2001b). Here, it would suffice to say that privatization implies a bidding process, including the monetary evaluation of the SEB. But this is a technical impossibility in an SEB for a number of reasons. For one information is structurally missing. Also the control rights to be valued by the investor have never been enjoyed by the SEB and remain theoretical. The privatization of an SEB can lead to underbidding if each investor takes a premium.

The above can be detailed on a simple model. Given non-integrated information and absence of delegation of power, the value

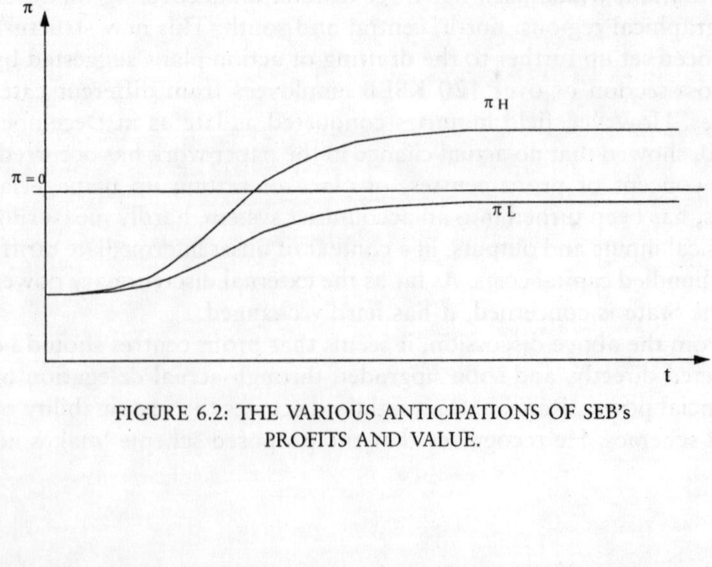

FIGURE 6.2: THE VARIOUS ANTICIPATIONS OF SEB's PROFITS AND VALUE.

of the SEB is unclear. Its value will ultimately depend on its anticipated profits, and these, on the basis of available reforms are constrained by the lack of information and delegation of power. Let us consider two external graphs for profits, one is high (H) the other is low (L) respectively, corresponding to the value $V_{0,H}$ and to $V_{0,L}$.

Considering the huge underbidding risk, we need to be cautious yet pragmatic. After all, the managerial capabilities of the private are even more in demand than its money. Let us then examine management contracts under this context.

6.3.1 Management Contracts

What do management contracts exactly recover and how could they be implemented in India in a credible and sustainable manner? According to Galal, Shirley et al. these contracts 'define the relationships between the government and a private firm contracted to manage a state-owned enterprise'. They further argue that they 'have generally been successful where they were attempted'. As pointed out by Girod, management contracts have succeeded in Africa, for instance in Mali. There is also the successful case of the Ivory Coast, where franchize[6] has occurred under the context of the clarification of the managerial rules and the transfer of the agents of the utility.

A management contract is implemented through a fee, 'usually a success fee but sometimes for a fixed fee as well', handed over to the private contractor. Galal, Shirley et al. specify that 'under this definition, the government does not receive a fixed rent (as it would be in a lease), it is responsible for fixed investments (which it would not be with a concession), and it holds majority ownership, ceding control only via an explicit contract (as distinct from a joint venture in which the government owns only a minority)'. According to Ranganathan, the fee is 'in return for specified improvement in certain identified parameters like system losses, revenue collection, etc'.

The two practical issues in the Indian case are the credibility of the setting up of control rights through a management contract, and the measure of its efficiency. Relative to the credibility, as far as the control rights are concerned, the experience of the reform of the distribution of electricity in Orissa shows that the control rights over manpower is crucial. Indeed, a management contract failed there in 1996-7. Rajan details that 'Gridco entered into a management contract called the distribution operations agreement, with Bombay Suburban Electricity Supply . . . to take over the power distribution in the Central zone, with effect from 1 October 1996. According to this arrangement, the private utility was responsible for distribution of energy, maintenance of the distribution system, and collection of electricity dues in the central zone. Though the initial distribution agreement was for a period of three years from October 1996, Gridco cancelled the agreement in April 1997, as it claimed that the private utility had shown a negative incentive of Rs. 0.28 crore over the 6 month period between October 1996 and March 1997.' On this failure, Rajan points out 'drawbacks in certain contractual provisions in the distribution operations agreement', particularly that 'the private utility had to manage the distribution business over 8,000 employees and engineers over whom it had no administrative control'. Lessons can be derived from this attempt. First, while the government might take charge of the payment of present employees,[7] a management contract should be designed with provisions to administratively sanction the faults, at least by posting different people in place of corrupt officers. Similarly, the contractor should be given the right to add privately-financed posts in the hierarchy. These would gather together some functions formerly given to the public officers, and being used along with the right to propose and finance reward schemes.[8] The contractor should also be given the right to invite the public officers to join these privately-financed type of posts.

As far as control rights on investment are concerned, although they are under the responsibility of the government under a management contract, the private contractor should be granted specific rights. Indeed, granting of full discretion over a given amount of investment would facilitate the implementation of urgent (and often very profitable) measures in a decentralized and prompt manner. The remaining part of the investment can be left to the government and be based on a process of negotiation taking place along specified rules.

As far as the cash flow rights of the private contractor are concerned, the fee should be both based on a fixed part (to sustain the commitment of the government) as well as a success-based one. Here the issue of measuring the efficiency comes into play. Galal and Shirley indicate that financial performance, productivity, and savings-investment deficit can be thought of as indicators of success or failure of reform. Given the non-integration of the reporting system in SEBs, and given the fact that the working force enters in productivity issues, productivity can hardly be thought of as a parameter in the fee, at least in the preliminary years. Also, if a management contract gives cash flow rights to the private contractor, it will a priori hardly give cash rights to the organization that it manages. Moreover, residual decisions on investment will not depend on the private contractor. In that case the savings-investment deficit cannot serve as an indicator. The financial performance remains a good candidate for this, and can easily be implemented through a way suggested by Galal and Shirley, that is, the operating surplus to sales in current prices. In utilities as well, it can be done through the increase of billing and collecting. Going according to this indicator, the related part of the fee should be success-based. This would allow the private contractor to fully establish an internal delegation of control rights based on an incentive system, and hence to fully use the provisions on control rights discussed earlier. A part of the management contract shall include provisions for setting up a proper information system, including measurement, metering and so on. The fixed part of the fee would relate to this role of the private contractor.[9]

Under this framework, an element of credibility can be ensured through long term contracts and explicit details of the mechanisms of conflict resolution. Indeed, Galal and Shirley underline that the 'Government and management must be able to demonstrate commitment. Governments can signal their commitment by agreeing to longer contracts and agreeing to explicit conflict resolution mechanisms, including international arbitration. For its part, the management can be seen as committed when the contractor has a reputation to worry about.' This reputation issue is valid for a government as well. Indeed, after a breakdown of relationship with a well-reputed contractor, the concerned government may find that its position is no longer credible toward funding agencies.

To sum up, it is proposed to:

— make provisions to administratively penalize the faults, at least by posting different people in place of corrupt officers,
— give the right and means to add privately financed posts in the hierarchy, with the right to invite the public officers to join these privately financed positions,
— grant full discretion over a certain amount of investment,
— base the fee both on a fixed part and on a success-based part linked via the operating surplus to sales in current prices.

With such specific provisions, a scheme based on management contracts has the advantage of establishing control rights whose credibility will rely on a reputation game, and proper incentives once these rights are set up. It also creates cash flow rights, at least for the contractor, if not for the organization itself. This also helps the management to benefit from the setting up of cash flow rights. However, the management contract will not a priori result in the setting up of cash flow rights for the SEB as an entity, since this is not needed in the design of such a scheme. This point may be reached only under some more or less degraded version of a privatization (leasing, concession) or through a full privatization.

In the latter case, if the government is willing to benefit from other advantages of these forms of private participation, then the management contract can be seen as a preparatory step. It allows the setting up of some of the property rights that will be evaluated under the bidding process. Similarly, the creation or enhancement of an information system limits the risk of inefficient selection linked with the information asymmetries among potential investors. But it must be underlined that it also remains conceptually separate from any privatization scheme. That way, a management contract may smoothen the bottlenecks of a direct privatization. This balances the point raised by Galal and Shirley, who maintain that 'under the conditions where management contracts can work, privatisation will often offer governments higher benefits (revenue from sale) and lower costs (no need to monitor, enforce and renegotiate the contract)'. Although this stands true if one considers the financial transfers at the moment of taking the decision, it is wrong when one considers the discounted benefit of a higher selling after some years of management contract. Having discussed the technical feasibility of a management contract, let us now discuss it with respect to the model previously presented.

6.3.2 Credibility/Sustainability: Management Contract as a Way to Build Information

Galal and Shirley, or Ménard and Shirley emphasize that all 'desirability', 'feasibility' and 'credibility' (or 'sustainability') matter in reforms. The desirability simply comes from both the public failure and the underbidding risk attached to straight privatization. The previous subsection has discussed the technical feasibility and the main conditions of a management contract. The point here is to discuss the advantages of management contracts compared to MoUs and decentralization schemes in terms of credibility/sustainability. The issue is not so much to ensure that the attempt is 'serious and fair', as to make it sustainable through in-built mechanisms.

The issue is then to ensure that the choice will be based on welfare issues, that is, to ensure that, once the decision to focus on efficiency has been taken by the state, its politicians will not renege, either fully or through ongoing executive instructions. In other words, the issue will be to maintain the suppression of the discretionary power of the state concerned. As far as this is the problem in hand, let us assume that the private contractor is able to discern whether he suffers executive instructions or not, i.e. whether the state exerts discretionary power on the matters within the contracted fields. Then, the implementation of the management contract can be modelled as such:

— As long as there is no problem, the management contract lasts. In terms of the previous model, the profits are on the (H) profile. Indeed, once executive instructions are suppressed and property rights are set up, then the further step of the reform process will start (elimination of the paper work, integration of an information system, replacement of procedures with cost-benefit analysis). This will arise because, in the proposed scheme, the private contractor will be granted sufficient incentives.

— Whenever executive instructions arise, it can be imagined that a negotiation process occurs, followed by a dispute settlement process if the dispute endures. In that case, if monitored through recourse to a multilateral or an Indian funding agency, the problem is not only advertised, but also leads very soon to the rupture of the management contract if it lasts.

Indeed, let us examine what happens if a management contract lasts. Let us suppose that the private contractor remains for as long as he enjoys actual managerial control. Under this scheme, the profits are on the (H) profile as long as the management contract lasts, and vice versa. That is, the sole fact that the investor remains, is the indication that the reform is on a good path, even if some time will be needed for putting in place an actual integrated information system and practical delegation of power. The ongoing nature of the management contract serves as an indicator of the followed

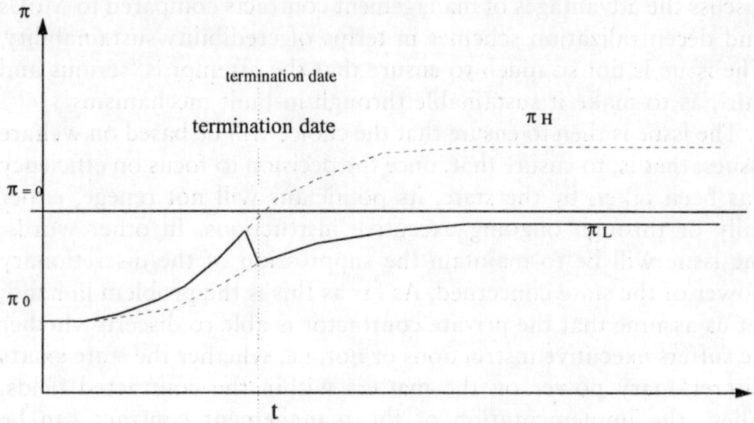

FIGURE 6.3: FAILURE OF A MANAGEMENT CONTRACT AFTER A PERIOD t.

path and provides information for all observers. Hence, success of reform is made visible under this scheme.

It might be useful to see how a failure materializes under this scheme, after a given period t. For that, let us consider that the transitory period of dispute settlement is short, and the information on the state of the SEB and the status of government interference is unclear. After the dispute, it is obvious that lock-ins may remain. Looking at Figure 6.3 it is clear that if executive instructions are issued, the SEB will be even on the (L) profile again.

Thus, the threat of the contractor to the concerned state is credible. Conversely, in case of faultiness of the contractor, the monitoring provides that the advertizing on reputation is strong enough to make the reverse threat credible as well. For the supporting agency, this also provides external information on the commitment of the state. To sum up, when working with a less binary model the probability of remaining on a high path increases for as long as the contract lasts. Whenever it stops, there is strong support for the belief that this probability is actually quite low. This is with that respect that the management contract can be a useful tool for the supporting agency in the monitoring of the reform process, and hence even for the state government in disbursement of loans by the agency. Thus, the management contract way looks practicable in Indian SEBs, both from the viewpoint of specific proposed arrangements as well as the information-building mechanism.

6.4 CONCLUSION

The set of organizational characteristics identified in SEBs (the lack of budgetary power and procedural rationality, decision-making not being based on costs, non-integrated information system, the general lack of measurement and proper accountancy) constrains the decision-making and leads to structural inefficiencies. These characteristics have been designed to cope with simple procedures aiming at developing the electricity system through the use of new equipment. This was consistent with the vision of the power sector in terms of quantitative growth and of a mono-technique (there is a mono-input—the planned expenditure, and a mono-output—the installed generation capacity), a vision suited to an early post-independence India where much electrification work remained to be done. But times have changed. Technical costs like maintenance of a developed network or supply of services specific to a diversified industry have risen manifold. In other words, this system is no longer adapted to a power sector that has reached a point of maturity, whose real need is a recurrent and decentralized cost-benefit comparison. The decisions based on procedures elaborated in an administrative context, and inspired by a mono-technique paradigm, now lead to productive inefficiency.

This chapter has suggested a scheme for the delegation of financial power. It has been argued that with a minimal set of arrangements regarding the monitoring of such a scheme, it can be started very quickly at the circle level. The expenses will be relatively small at the size of the board, but are guaranteed to yield high profitability and subsequent revenues at the circle level. Well monitored, a learning process can occur, to generalize the experiment to other parts of the budget.

It has been further argued here that the most credible environment to set up such a delegation of financial power is a management contract with a private partner. Based on the Indian experience, it details its principal characteristics for success, mostly insisting on guarantees on management, through an original proposal of a financing scheme for 'parallel hierarchy'. Given this guarantee, the contract provides to the supporting agency and to the electoral constituencies external information on the commitment of the state. If the former makes clear that it will support a management contract scheme implemented in that way, then the bias for opportunistic behaviour of the state is partly suppressed. In that respect, the management contract also offers a tool for close and reactive monitoring, thus creating an opening for phased disbursement of loans.

This scheme allows efficient reforms without necessarily selling assets to the private sector if such is the choice of the state or a constraint imposed by the market (in a privatization, though sellers are required, takers are all the more in demand, a phenomenon not at all evident in India). In the other case, if there is a political choice to ultimately privatize, the proposed mechanism is the sole one which offers to the investor the necessary information, and thus allows privatization in satisfactory conditions both in terms of economic efficiency and speed.

NOTES

1. Only small items can be actually purchased by issuing a check.
2. Beyond the analytical interest, this issue is a very practical one in some States facing a high level of transformer damage.
3. It can happen that, when some transformers get damaged, the population repairs them with some local devices that don't fit the technical norms, like by putting higher fuses for instance. In this case the field officers feel helpless in controlling them.
4. Indeed, overloading is high. Dispersing the investments might reduce the overloading duration locally, but in peak hours the problem will persist. Hence, burnings would go on at only a slightly reduced rate. Instead, treating some areas as priority would allow full suppression of burnings in these areas over period 1, the spare funds being invested in a second zone with concentrated investments over period 2.
5. Ranganathan's proposal is in that case designed for the Nepal Electricity Authority, but as far as organizational characteristics are concerned, they are very close to the situation in SEBs. Details can be found in V. Ranganathan, 'Commercialization of Nepal Electricity Authority', report for the Nepal Administrative Staff College, 50 pp., Indian Institue of Management, Bangalore, 1997, and in V. Ranganathan and Sanjay Gupta, 'SEB Performance: lessons from Thailand', *The Hindu Business Line*, 1 February 2000.
6. Compared to management contracts, franchize is a step further toward privatization, by allocating more control rights for a longer period. But it still keeps the property on assets with the state and allows the whole set of possibilities by the end of the period. Renegotiation, taking over, or privatization are the available options.
7. As well as taking care of the general wages revaluation in the public function.

8. This can even be financed through a mechanism of benefit appropriation, similar to the one proposed in Chapter 5.
9. This latter provision answers the reluctance shown in interviews by some reformers. They argued that a management contract, in case of SEBs, would have to address the fee issue in a context of unknown level of technical and non-technical losses. However, this is already an issue in the scheme of privatization, to be addressed by regulatory commissions in tariff setting. I argue that, in a context of a management contract, this is of less importance, since the owner is still the state, and therefore this uncertainty does not come into a bidding process. Also, the completion of the work of establishing an information system can be easily checked through objective and clear criteria, like installation of meters at different levels, restructuring of reporting and so forth.

REFERENCES

Galal, Ahmed and Mary Shirley et al., *Bureaucrats in Business*, World Bank Policy Research Report, New York: Oxford University Press, 1995.

Girod, Jacques , 'Les industries électriques du Burkina Faso, de la Côte d'Ivoire, du Mali et du Sénégal: performances et modèles organisationnels', *revue de l'énergie*, no. 448, Paris, 1993.

IMI, *A Study on Financial Management of the State Electricity Boards*, a three volume report for the Finance Wing of the Ministry of Power, International Management Institute, New Delhi, 1999.

Ménard, Yves and Mary Shirley, 'Reforming contractual arrangements: lessons from urban water systems in six developing countries', Working Paper of the World Bank Research Institute, 40pp., Washington, DC, 2000.

Rajan, Thillai, 'Restructuring of a Vertically Integrated Electric Utility: a process study', dissertation, fellow programme in management, Indian Institue of Management, Bangalore, 2000.

Ranganathan, V., 'Commercialization of Nepal Electricity Authority', report for the Nepal Administrative Staff College, 50 pp., Indian Institute of Management, Bangalore, 1997.

Ranganathan, V. and Sanjay Gupta, 'SEB Performance: lessons from Thailand', *The Hindu Business Line*, 1 February 2000.

Ruet, Joël, 'L'organisation du secteur électrique Ivoirien', mémoire de DEA, Université Paris-I Panthéon-Sorbonne, 1997.

———, 'Winners and Losers of the Reform of SEBs: An Organizational Study', Occasional Paper, New Delhi: CSH, 2001.

———, 'A new paradigm for the Indian power sector: going beyond the "silver bullet" privatization', paper presented at the Indo-French Conference on Privatization, Hyderabad, 6-8 March 2001, to be published in the *IPE Journal*, Hyderabad, 2001a.

7

Preparedness of State Electricity Board for Privatization

PUNEET CHITKARA, PREM K. KALRA
AND RAJIV SHEKHAR

Infrastructure facilities and services have been prime candidates for privatization for well-known reasons such as losses (financial, and technical) and inefficient operation. However, the recourse to privatization has generated a lot of backlash. The assumption that privatization of an organization would automatically enhance its operational efficiency and bring down the customers' bill has not been observed to work out in practice.[1] Thus a question naturally arises: Is our understanding of the privatization process correct?

A continuous increase in the demand-supply gap, high technical and non-technical losses, and poor revenue collection meant that the cash-strapped state governments did not have adequate resources to prop the ailing power sector. Consequently, governments looked towards the private sector to (i) fund capacity expansion and (ii) increase operational efficiency. On the face of it, the governments' reasoning seems credible. Unfortunately, privatization of the electricity sector has major political and social dimensions and cannot be viewed simply in terms of profits and losses, as with BALCO and Modern Foods. In fact, the government's efforts in privatizing the electricity sector have not yielded encouraging results. Some lessons can be learnt from the following past experiences:

- The first major mega power project in the private sector managed by Enron Corp. of USA has been grounded because of unusually high tariffs.
- Privatization of distribution in Orissa has not been successful, distribution companies in the state are facing serious liquida-

tion problems. Not surprisingly, no private bid came forth for acquiring distribution rights in Haryana and Kanpur.

— As a prelude to privatization, state governments are organizing the SEBs into three separate entities: (i) generation, (ii) transmission, and (iii) distribution. Unfortunately such classification has not improved power availability in UP where daily power cuts of 8 hours or more are common even in major cities like Kanpur. Moreover the record of the newly created UPPCL in collecting dues is not significantly better than its predecessor, the UPSEB.

The following questions naturally crop up:

— Why have the government's efforts in privatizing/corporatization of the electricity sector not borne fruit?
— Why do decision-makers think that the privatization is the only solution to the ills of SEBs?
— Did any SEB take steps towards privatization? If yes, were these steps appropriate for privatization?
— Did the decision-makers ever think that privatization could result in a backlash or create difficulties? If yes, where are the plans to implement such schemes?
— Political interference has played a major role in landing the SEBs in their present hopeless situation. Are we sure that privatization will minimize political interference?
— Further there are technical issues related to SEBs: Is it possible to reduce losses? What are the main reasons for losses? Is it theft or overloading of the system or poor maintenance of the system or excess subsidies?
— Will customers be happier and face less difficulties in resolving their problems after privatization?

7.1 IMPLICATIONS OF PRIVATIZATION

It is very important for a government that decides to privatize to understand the implications of privatization, primarily because the privatization of the power sector is different from privatization of hospitals, steel plants or schools. For example, consider the localities and industries, which are not paying bills today. Would they need to pay their dues in the post-privatization scenario? In such a situation, the operator has a choice either to cut services or hire middlemen to collect dues. Cutting services may not be easy because of considerable social and political pressures. This problem is further complicated by the fact that the electricity sector is interminably intertwined with the social and economic progress of the country.[2]

One has to understand that privatization is a movement and not simply a game of financial profit and loss. Moreover, significant efforts have not been made to identify reasons, barring corruption, for the poor functioning of management and financial losses in the electricity sector. Even today we lack micro level data, its analysis and understanding. It therefore becomes necessary and essential that a systematic scientific study beyond the paradigm of loss and profit, measurable in monetary terms be put forth to examine the preparedness of SEBs for unbundling, change of ownership, disinvestments, and privatization.

Thus a question naturally arises: is our understanding of the privatization process, especially in the electricity sector correct? If yes, we need to develop a systematic approach for privatization through minimization of risks involved and benchmarking. Identification of observable and controllable variables affecting privatization should be investigated formally and a case study may be generated to illustrate the methodology.

7.1.1 Risk Analysis

The project may turn out to be quite risky with respect to the risk variables taken all together. Overall the project is worthwhile to undertake but what needs to be kept in mind is that the control of investment costs, the willingness to pay estimates, and the growth of GDP are all critical variables determining its economic feasibility.

This paper seeks to develop a composite measure of risk involved in privatization of a distribution utility. Factors contributing to the total risk that an investor perceives while buying a distribution utility can broadly be classified as

— Inadequacy of existing Assets
— Revenue Risk
— Regulatory Risk

7.1.2 Inadequacy of Existing Assets

When a utility is privatized, both the physical and the human assets are transferred to the private company. The physical assets, such as Circuit Kilometres of Distribution network, auxiliary services and transformation capacity may be inadequate to serve the existing load and hence could be getting overloaded leading to high distribution losses. The private owner would then need to invest in capacity expansion after the distribution zone is privatized. The existing government-owned distribution systems are highly overstaffed and the workforce lacks the required motivation to work towards maximizing firm valuation goals, which the private owner would be interested in.

Inadequacy of existing assets may be measured in terms of:

- — Degree of lack of duplication for an electrical plant such as lines, cables, transformers, auxiliary equipment such as reactive devices, etc.
- — System upgradation through dynamic rating of lines, etc.
- — Deficit/surplus generation and transmission capacity available.
- — Electrical protection equipment used to disconnect a faulty portion of the system.
- — Ruggedness of design and construction standards.
- — How well the plant and equipment is maintained and operated.
- — Organization overhang in terms of excessive manpower.

7.1.3 Revenue Risk

Given the regulation, this depends on the zoning of the distribution network. The state government chooses between a concentrated zone and a mixed zone model. Because of the load profiles, there are different risks associated with each of these zones. These risks get exacerbated due to:

- — Inadequacy of metering
- — Billing Inefficiency
- — Collection Inefficiency

It may be argued that the very purpose of privatization is to improve the above-mentioned inefficiencies and hence the risks on these accounts may not be allowed to affect the allowed returns to the privatized distribution utility and hence the credit rating of a distribution zone. But it has been experienced that overcoming these inefficiencies requires time in terms of installation of meters (which may not only be difficult but also impossible under certain circumstances). Also, since the manpower is transferred as such, the private utilities would take some time to change their mindset through appropriate HR policies.

7.1.4 Regulatory Risk

A private sector participant looks for certainty of the treatment of all major elements of revenues and expenditure (including losses), which assures him of being able to project profitability reasonably before bidding. For all practical purposes, the interest of private sector in the restructured entities for distribution would be adequate only if the new companies seem to break even within a short time frame of a year. Retail tariffs are regulated. Hence the cash flow of the distribution utility is subject to a lot of regulatory risk. In the determination of tariff for retail consumers, there are these risks:

— Determination of the value of assets at the time of transfer, keeping in mind the inadequacies listed under 'Inadequacy of Existing Assets'. The extent of consumer contribution to bridge the inadequacies [i.e. percentage of contribution towards construction (or system strengthening) work in progress] till the system performance reaches a minimum acceptable level needs to be clearly defined. Not all existing distribution utilities maintain fixed asset registers. Hence it becomes difficult to appropriately determine the cash flow on account of depreciation.

— Most of the distribution companies have adverse debt coverage ratios and these liabilities would need to be restructured before the assets are transferred. For example to cover the debt liabilities of DVB, it would require 10 annual installments of Rs. 700 crore each.

— For most of the distribution utilities capitalizing certain expenses to project enhanced profitability has artificially deflated total expenses. Such expenses need to be uncovered to the extent possible.

— The state regulatory commission, following the policies of the state government, fixes the retail tariffs. The past record of the state government in making good the subsidies provided to the State Electricity Boards needs to be taken into account to credit rate the distribution assets.

— Technical and non-technical losses can be reduced in a phased manner after privatization. The rate of reduction in losses needs to be determined based on the best industry practice using appropriate benchmarking techniques.

— Appropriate load forecasting techniques for each consumer category and both for real and reactive power requirements need to be used for determining the additional capacity expansion requirements and hence appropriately determine tariffs (share of consumers in the expenditure towards construction work in progress must be decided).

— Financial mechanisms for treatment of un-metered power (such as municipality loads) should be put in place until appropriate metering devices are put in place. Equity betas have been estimated to be between 0.95 and 1.33 internationally. These could be adjusted for country risk and state risk for arriving at the rate of return.

— Tariff Revisions: In order to provide incentive for efficiency improvement the state regulatory commissions need to put into place multi-year tariffs. The yearly adjustments can be on the basis of RPI-X mechanism, instead of just allowing for a pass through of all costs.

7.2 BENCHMARKING

7.2.1 Objectives

Benchmarking will provide information about SEBs:

— Status of finances
— Technical health
— Quality of personnel
— Proper evaluation for its price
— Life and future performance

A comparison of power quality at consumer level required across all states could be materialized provided an effective criterion for benchmarking the power sector is set-up. This would also act as a pointer to the quality of life of the consumer and can be even used for predicting future demands.

7.2.2 Methodology

Attribute Identification: Identification of appropriate attributes is an important pre-requisite for benchmarking. For example, the conventional method for characterizing power quality in terms of voltage and frequency may not be sufficient. With increasing use of computers and automated appliances, harmonic distortion should be an additional attribute for benchmarking.

Attributes may also include preferences of a particular state due to its location, employment patterns, average earning capacity, etc. One would also face situations where some kind of equivalence has to be drawn, for example agriculture is important in Haryana and Punjab while in Bihar the mining sector is the bread earner; hence the influence of the bread earning sector on the electricity sector requires to be accommodated.

Sensitivity Analysis: Sensitivity analysis exposes the important variables that affect the final outcome of the project. The results of the sensitivity analysis often signal the variables that must be considered in the risk analysis. They also tell us the important policy variables that the management should monitor and evaluate carefully during project implementation. For example, the sensitivity analysis of the project may project that tariff policy is crucial to its financial health. Further it can also bring home that inflation, total connected load and taxation have a direct implication on returns to the company.

Techniques: There are formal methodologies to evaluate and measure the performance of a large and interconnected industry. The techniques, which are becoming popular include Data Mining (DM), Data Envelopment Analysis (DEA) and Neural Networks (NN).

In DM, features are extracted to build clusters indicating the ratings. This provides us benefits of handling a large database and building relations among various attributes to compute the performance of the service. For example, we can consider the reactive power requirement on an hourly basis and harmonic pollution on an hourly basis for the last 5 years. Data mining methodology would help a user to find the patterns of reactive power over days, weeks, months, and years. This information can be used to cluster the relationship about reactive power consumption with mechanisms used to alleviate problems created, such as low voltage. Similar concepts can be used for demand management.

DEA utilizes techniques such as mathematical programming, which can handle a large number of variables and constraints. DEA can be used to evaluate subsystems and integrate the performance of subsystems to determine the performance of the complete system.

DEA analysis[3] was used to examine the performance of 25 Japanese companies whose business environment is now undergoing a dramatic shift from governmental regulation to competition. Results showed that the wholesale generation (WG) companies have operated more efficiently than vertically integrated and investor-owned (VIIO) companies. An important policy implication for electric power generation emerges from this study. Integration between generation and transmission, as in VIIO companies, may not enhance (production-based) efficiency.

Neural network can be used to predict the reactive power requirement on an hourly basis and hence actions can be planned to encounter the expected problems.

7.2.3 Benchmarking is Dynamic

Dynamic: Power quality has deteriorated quickly in the last 4-5 years due to changes in the nature of load. This change has taken place due to computerization, automation and mechanization. In the past, only voltage and frequency variations were of concern. Now computers are generating harmonics in excess; apart from the UPS, switching appliances like air-conditioners, refrigerators, and TVs have also contributed to a change in the nature of load. Thus, instead of power factor, displacement power factor has become more relevant today.

Deliverables: The deliverable of benchmarking is the development of local performance indices (LPI) to benchmark each activity in the organization. Finally, the LPIs should be combined to arrive at a global performance index (GPI), akin to a credential rating, for each SEB. For example, LPIs for generation may be the residual life of a generating unit, environmental impact, nature of combustion technology, and maintenance costs. Similarly, LPIs for distribution may be losses, reactive power requirement, connected load, projected connected load, life of major equipment such as transformers, power factor and/or displacement factor. Line availability, overload capacity, compensation mechanisms, and voltage levels as well as losses would constitute the LPIs of the transmission system. It needs to be pointed out that GPIs by themselves may not convey the correct picture of a large, interconnected system. For example, an efficient generation and transmission system may be weighed down by poor distribution.

Considerable amount of work is required to clearly bring out the procedure of benchmarking. This would also assist in comparison of the various SEBs. It is also well known that investors carry out their studies before they buy. A benchmarking procedure would be an ideal tool to examine their understanding about the SEBs.

This study must include the development of tools for the benchmarking, and evaluations of technical and financial health. One could even develop a procedure to provide credit ratings/credential ratings to SEBs.

7.3 A PROPOSED MECHANISM FOR PRIVATIZATION

Privatization involves three parties, i.e. government/state, private buyer and customer (user of service). Even though customers are not directly involved in the negotiations between the buyer and seller, their behavioural pattern has a definite influence on the bidding process. For example, clusters consisting of disciplined consumers would be easily sold compared to clusters having a large proportion of 'free loaders'. Consequently, to make privatization successful on a global scale, it is imperative to instill confidence in the customers at large.

SEBs can be unbundled into four services:

— Generation
— Transmission
— Distribution
— Auxiliary Services

In our view, the natural choice of initiating privatization is to start with privatization of auxiliary services, primarily for the following reasons:

— Total investment made by the private investor would have been minimum.
— Improvement of voltage profile and increased power availability can build customer confidence for privatization. Farmers have often demanded greater availability of power, even at a cost, instead of irregular and poor quality free power.
— Transition of SEBs to the market environment would be gradual. Moreover, introduction of auxiliary services would in any case improve the system strength.
— Once system strength is improved the privatization would be enhanced.

Benchmarking of auxiliary services is not a difficult task. Auxiliary services can be treated in a manner similar to generation. The only difference is that in generation the primary objective is to meet real power demand whereas auxiliary services would provide support to reactive power and filter for harmonics. These services can easily enhance system capability up to 20 to 30 per cent. Such an approach would also send signals to the common man that we are thinking of improving their life rather than snatching away his free lunch.

7.4 CONCLUSION

The thoughts presented in this paper might be tentative but hopefully they manage to present an innovative approach to (i) evaluate the preparedness of SEBs for privatization and (ii) the privatization of the electricity sector. It is recommended that a project be funded to crystallize the concepts presented in this paper. The deliverable of the proposed project would be a case study to illustrate new methodologies for developing a GPI and hence the credential rating of SEBs. The credential rating would assist investors as well as the government to make an assessment of the health of SEBs with the minimization of various risks associated with the process of privatization. This mechanism of generating a credential rating would also streamline privatization. Further, it has been suggested that the confidence of the users of the electricity sector be enhanced to build an atmosphere conducive to privatization. We strongly believe that auxiliary services should be privatized first. Also the concept of power quality must be finalized in the Indian context.

NOTES

1. Distribution privatization in Orissa is a case in point.
2. The electricity sector affects almost all walks of life ranging from education, health, water supply, transportation, agriculture, etc.
3. T. Sueyoshi and M. Goto, 'Slack-adjusted DEA for time series analysis: Performance measurement of Japanese electric power generation in 1984-93', *European Journal of Operational Research*, vol. 133, 2001, pp. 232-59.

8

Restructuring Gujarat Electricity Board: Outline of a Strategy and Proposal for Action

SEBASTIAN MORRIS

This chapter is divided into five sections. The first section reveals the reasons for urgent major reform of the electricity sector and the restructuring of the GEB. In Section 2 the strategy is outlined. Sections 3, 4, and 5 respectively bring out the key proposals, discuss the modalities and lay down the scheme in a sequence of events and decisions.

8.1 THE NEED FOR REFORMS INCLUDING SUBSTANTIAL RESTRUCTURING

The privatization process in the power industry that began in 1991-2 has severe limits as far as improvement in efficiency is concerned. They have also made the task of further privatization difficult. This is because the reforms have not so far addressed the issue of tariffs and realizations squarely. With continuing problems at the distribution end, where the gap between the electricity sold and accounted for is continuously widening, there is not enough cash flow to justify IPPs. Moreover, the IPPs see high payment risk, which inter alia makes IPP power costly. Considerable, if not total, shifting of business risk onto the state system has taken place in IPP contracts. Limits to escrowability have been reached and the liabilities of the state would further increase as the IPPs come on stream.

There is massive revenue loss from the GEB on account of high T&D losses both technical and commercial (due to theft and non-payment), auxiliary consumption (whether in fact, or due to theft) and over-reporting of agricultural consumption.

The demand for budgetary support to cover not only the true subsidy—due to government policy to subsidize agriculturists—but also avoidable leakages and losses, is large and increasing.

Tariffs for industry have already reached unsustainable and dysfunctional levels. GEB rates are higher than the cost of self-generation even at small size (~5 MW), and within the constraints of the present captive generation policy. It is only the current high interest cost of borrowed funds and generally recessionary conditions that prevent an exodus of bulk consumers.

High tariffs affect the competitiveness of Indian industry. Acting through Gujarat's specialization in power using and power critical industries, they also impose special discrimination against Gujarat as an industrial location.

Low tariffs for agriculture, and more importantly unmetered supply have resulted in excessive and wasteful consumption, and consumption for unauthorized uses, besides creating the basis for moral hazard. The agency (GEB) that sells is the same as the one that reports the identity of the consumer to the principal (the government). Since the government lacks the ability to verify the identity of the subsidized consumer, and the amount actually consumed by such consumers, exploiting the agency problem becomes a focus of the agent (GEB).

The state electricity system is not viable and would soon collapse financially unless set right. Not only the viability of the SEB, but the survival and growth of private provisioning is at stake. Reforms that correct the key agency problem of the GEB, as outlined above are vitally necessary.

No piecemeal or stop-gap arrangement, that allows the continuation of the agency problem and does not rein in the leakages, is possible. Thus, proposals to sell generation assets of the GEB, without a complete reform of distribution are not meaningful.

The dead weight losses which are true social losses, and not just a transfer problem, are very large. They take the form of losses due to unutilized and underutilized capacity, both within the system (GEB) and outside (other power suppliers and captive generators). Allocative efficiency loss resulting from inappropriate technology choice, violation of merit order, disincentives that prevent demand from responding appropriately to prices, are even more severe.

The independent regulator, viz., the Gujarat Electricity Regulatory Commission (GERC) is already in place, and is working. GEB has asked for a tariff revision that is upwards of 20 per cent. Since the GERC is independent and is empowered under the law to protect the interests of the consumer, it was in a position to put the burden on GEB, by pointing out the vast leakages. It has ruled a tariff increase, that was not substantial enough to cover even the post subsidy deficit of the GEB, asking it to further reduce leakage instead of relying on a tariff hike.

The issue of capping the subsidy, and separation of the administration of subsidy from the power entities are important. Thus, stopping the leakages and improving efficiency can no longer be ignored.

8.2 THE STRATEGY OF RESTRUCTURING

We may define reform as a kind of reorganization bringing in legal and policy changes, which would:

(a) solve the agency and moral hazard problem in the administration of subsidies;

(b) create incentives for cost saving and efficient generation, transmission and distribution;

(c) correct the inversions in the tariff, and lead to a scheme that would allow prices to play their economic role of reflecting costs, and being a signal for outputs to respond.

No reform is possible unless the administration of subsidy as such is entirely separated from the entities (whether state-owned or private) that manage and operate the electricity business.

Separation of subsidy administration alone would not be sufficient to solve the agency problem. In order to neutralize the vested interests, both within and outside the GEB, that have come up around the 'leakages' of revenue, it is necessary that distribution be privatized. The culture of tampering with meters, coming to an agreement with the customer, etc., is too well-entrenched to be removable by mere administrative, financial or formal restructuring, such as corporatizing the GEB.

Privatization of distribution that leads to efficiency, and elimination of revenue leakage would result in commercial viability of all backward linkages including generation, transmission and fuel supply. This would allow for these to be run on commercial principles, privately or otherwise, and hence for the sustainable and autonomous development of the sector, as a whole. Budgetary support, except to the extent of a capped subsidy, would not be required and could in fact be brought down.

Separation of subsidy from the distribution entity would make it possible to cap and direct the subsidy, so that the distortions are removed and the social returns to the subsidy are maximized.

Quick privatization of distribution, even before regulatory clarity and strategy emerges, is the need of the hour since with every month the leakages and continuing inefficiency accumulate, imposing a severe burden on the state budget. At a rough and conservative estimate the leakages alone every month would exceed Rs. 80 crore.

Such quick, and revenue enhancing privatization, that at the same time does not retard the scope for further reform and change, would call for bundling of some generation assets with distribution. This would simultaneously lower the demand risk for generation assets, allowing them to be privatized easily (without creating PPAs, contingent liabilities, etc.) and simultaneously lower the supply risk for distribution entities.

Thereby, and to the extent that the assets are bundled, their value is not dependent crucially on the ability of the restructuring and regulatory processes to unbundle and put the system back together through elaborate market-creating processes. Thus, the regulatory and 'restructuring' risk can be minimized for bundled entities. Resources can then flow from the capital markets, if the final tariff awards are consistent with recovery of costs (even if they are in error with regard to recovery of particular costs). Thus, the regulator would also have the space to organize and lay out his strategy, and with much caution, and to implement the same in a phased manner.

Current capacity within the GEB (both together in the IPPs, in captive generators and in Gujarat's share from the central electricity companies) is underutilized. There is much scope to shift demand to off-peak hours to raise the system-wide PLF higher and improve power quality. Thus, more efficient capacity utilization possible with a workable time of the day tariff (leading to shifting of demand) is not feasible unless distribution and generation company staff have the requisite freedom to offer substantial discounts that depend upon the situation. Such freedom in government departments and public sector companies, given their rule orientation, would be difficult if not impossible. Functional (rather than the usual formal) monitoring of such discounts would be difficult again, unless the entity is privately managed. The scope for shifting to off-peak hours is large in Gujarat given the fact that Gujarat has a large number of power-based industries, and a substantial agricultural demand, which could use off-peak power effectively. Most importantly, bringing back accountability and embedding responsibility in the present GEB without change in ownership would be very difficult.

In moving towards a restructured system where privatization of distribution and generation is the crucial aspect, the processes followed to bring about these changes are important. Success would depend upon two critical aspects: It would be necessary to build in complete transparency and information sharing with bidders, consultants, study groups, concerned departments of governments, power users, captive generators and citizens in general. Similarly, the restructuring process would have to be directed by the appointment of a credible person who not only has.the necessary skills, but also the independence from GEB, the government, influential politicians, the bidders, consultants and advisers. This is vitally necessary since the incumbent management and staff of the GEB, the government or influential politicians and bureaucrats, cannot be assumed to be uninterested parties.

Equally important, complete transparency would result in the best prices from bidders for GEB's businesses and assets. The true status of GEB's assets is hardly known and only complete transparency can ensure that bidders do not bring legal claims after takeovers/assets sales. Transparency and the fairness that can come only with independence and expertise would also attract many more bidders, than only those who seek to gain out of governmental failures, or out of their ability to unduly influence the government.

Transmission ought not to be fully privatized. This is to ensure that the option of going for a complete market solution in generation through pool prices, contracts, etc., is kept open and becomes easier to pursue. It avoids the danger of 'artificial' constraints in transmission segments, that give rise to shortages in particular territories, which in turn result in high-realized prices for generators.

Yet, since good transmission performance is always in the interest of the consumer the transmission entity should have substantial participation from consumers, including user industries, while being entirely free of shareholding by generation and distribution interests.

Absolutely central to enhanced efficiency of the restructured system is the degree of competition among generators.

8.3 THE PROPOSALS IN BRIEF

Existing transmission assets of the GEB would be formed into a company (TRANSCO) with the Government of Gujarat holding 50 per cent and the remaining held by consumer groups and the general public. Distribution-cum-generating companies (DCGs), generating companies (GCs) and captive generating companies (CPs) would not/should not be allowed shareholding in the TRANSCO.

TRANSCO should lay down detailed guidelines, announce connection, congestion and line-loss charges for various types and sizes of loads connected to the transmission network and for use of its facilities, in consultation with the regulator. The assets would have to be valued at replacement cost minus the maintenance costs required to bring them to a state of complete repair. The initial price for services through a price cap, or the return to the transmission company at 16 per cent or more would have to be fixed by the regulator.

TRANSCO should not carry out any power trading on its own account, but should be responsible for system integrity and lay down appropriate rules for optimal operation of the system that is closest to merit order. All DCGs, GCs, IPPs and CPs will have equal access to the grid. TRANSCO should, along with the regulator, facilitate and lay out the rules for the development of a wholesale market.

Surpluses, in the sale of generation and distribution assets of the GEB, and the current capital contributions of the Gujarat Government if any, will go towards enhancing TRANSCO's network and quality of assets as an important priority, since 'extra' transmission capacity would be necessary for ensuring competition among DCGs and GCs.

With the government holding less than majority stake, the company would not be 'state' in the courts' view and would have the requisite freedom to pay its staff well enough and in a manner that creates incentives within the organization for performance. Similarly, it would not be limited by the current constraint of being 'state' in labour management in general.

Privatizing the existing distribution zones would not pose any problem when the subsidization is direct. As such, each DCG would, for its distribution business, have realizations per kWh of electricity sold that are very nearly the same. Yet, the costs incurred in distribution to urban and bulk consumers would be lower than for rural and small consumers. Except the Bhuj circle, the variation in distribution costs, while significant, would not amount to a basic bottleneck against privatization. If the regulator adopts ROI type regulation, it would mean somewhat different ceiling tariffs to the final consumer. The differences would reflect the varying average cost of distribution alone, arising out of the variation in consumer mix across the zones. This is likely to result in a variation no greater than about 20 per cent in the final price of power to the small consumer. On a price cap based regulation, it would mean different (by about 20 per cent at most) consumer end price-caps, for the zones to start with. In the long run the distribution part of the price-cap would be larger in the more rural and dispersed zones.

To arrive at the base cost for distribution, its indexation, or the return, a detailed study of distribution cost (variable and fixed) as a function of the various measures of physical network the density and type of consumers, would be necessary. With ROI type regulation, an announcement that the DCGs would on their replacement value of assets earn a return of 16 per cent, would remove any regulatory risk in the sale of both generation and distribution assets. It would then be the regulator's annual task to approve the tariff demands placed upon it by the DCGs. With price-cap regulation, the regulator would have to announce his formulae of prices for various cost segments.

Even if the regulatory proposals are not in place in their details at the time of restructuring, the models in approving tariffs for the present licenses (AEC, SEC) could be extended to the new DCGs by the GERC. These would come with appropriate incentives to control distribution costs, improve efficiency and service quality. The best choice for the regulator would of course be a price-cap approach to setting prices for different components of transmission and distribution costs; fixed, variable and penalties in exceeding designated loads.

In bundling the particular generation assets to go with the distribution zones, i.e. in the construction of the DCGs, the first consideration would be to ensure broad similarity in terms of the distribution of capacity over variable costs. The second consideration would be the 'nearness' to the distribution territory. With appropriate bundling, a little less than half of the average output of the DCGs could be internally generated. For the rest they would buy power from the generating companies and captive generators, through standard negotiated contracts, where the price would vary depending upon the demand and supply position. The nature of the demand (to degree to which interruptible) would also inter alia influence the price.

Increasing demand would ensure that the capacities of most GCs as also the generation assets of the DCGs and CPs would be used at rates on the whole better than at present. Similarly, the complete removal[1] of all current distortions in the bulk production and sale of power would lower the cost of bulk power below current rates to reflect true costs. The quantum of power could increase too, as all the capacity, including that of captive generators would come on stream during the peak hours in response to high prices during peak hours for bulk power, when the DCGs would be important buyers.

Allowing all loads at 1 MW or above and all CPs with installed capacity 1 MW or above, the right of access to the transmission/ distribution lines would be the key to bringing about competitive pressure. For such customers/ entities, the sale and purchase of power is completely separated from its carriage. For them the transmission/ distribution company including DCGs operate as common carriers (which are regulated).

The rules for market making would have to be spelt out by the TRANSCO, after system and market simulation studies, that would alert it to possible pitfalls. But in the meantime simple rules that insist upon matched contracts for sale and supply with less than 5 per cent variation, by the TRANSCO, before allowing use of the grid by all participants, would serve the purpose. They would have to be checked for systemic feasibility given the grid capacities through a simple programme.

Hydro-power stations should be bundled together and sold as a separate generating company (HYGEN). The advantage in such a company is that it could specialize in offering power that caters to unforeseen demand. During monsoons when run-off is available, HYGEN would have an incentive to lower its price considerably. This would make coal-using base stations schedule their annual planned maintenance during this period. HYGEN would then develop as a specialist peak power player and lead the market developments in peaking power.

IPPs hold contracts with the state through their PPAs which allow them a stream of 'certain' revenue flows, as long they are able to show contracted availability and meet standard, agreed norms for heat rate, etc. Indexation of the price of electricity purchase to fuel prices, have transferred both demand risk, fuel price and exchange rate risk onto the SEB (state). Therefore, change towards a market model would be naturally opposed by the IPPs, if their interests are not 'protected'. The original policy under which the IPPs came about is the guaranteed 16 per cent return at 68.5 per cent PLF. Returns above these are specific to the project and the government is under no legal requirement to protect the same. These are essentially returns to 'regulatory speculation' and to 'efficiency'. As such only the 16 per cent on equity share capital needs to be treated as arising out of stranded contracts and provided for.

Several options are possible. One would be to buy out the IPP's equity at a value equal to the capitalized value of a stream of returns at 16 per cent for twenty years (the life of the project). Thus for Re. 1 of equity the cost would be Rs. 1.21 approximately, at a discount rate of 12 per cent.

IPP costs per MW are higher than current costs (both because of padding of costs, and fall in turbine and equipment costs). Until the time the IPPs reach financial closure, the net cost of buying up IPPs and starting on a clean state is likely to be between 0.8 crore to 1.3 crore/MW. Thus, if economic pricing of electricity is to be the norm, this loss would have to be covered by the government, unless it wants to renege on IPP contracts. Therefore, a certain provision for 'stranded contracts' would have to be made.

It would be best to provide for this uniformly across all consumers as a separate charge/kWh of electricity consumed, and the fund so created be used to service the financial cost of buying up IPPs at between 0.8 crore to 1.3 crore/MW.

This fund would also have an inflow from the BPSA contracts at a rate of approximately Re. 1.0 (difference between average contract price under various BPSA (Rs. 1.38 ~ 1.50) and long run marginal cost of bulk power at Rs. 2.5/kWh. Thus, the actual credit on this account per year would be 1293 MW (GEB share of CPCs capacity) *0.75 (PLF)* (1-0.15) (distribution losses) = Rs. 849 crore per year. Assuming the BPSAs have an average residual life of five years this cash flow would have a present value of approximately Rs. 3,000 crore. This would allow the government (the holder of these lucrative contracts) to buy up between 2,300 MW to 3,800 MW of IPP capacity, if the appropriate SPV for restructuring can be created.

Thus, the increase in tariffs that would have to be provided for for the 'stranded contracts' may not be particularly large, if the value of the BPSAs can be nearly fully internalized.

Even before reform along the lines in this note is agreed upon, all further negotiations for IPPs, and all pending negotiations, not yet formally contracted would have to be stopped from this point on.

Alternatively an SPV to recover stranded contracts could be set up, which would buy power from IPPs and CPCs, and pool and sell the same to DCGs and bulk buyers. The SPV needs to be constituted so that it has a life of twenty years, is completely transparent, and entirely open to the regulator and the public. It would have the mandate to sell power at the lowest feasible cost to DCGs and others, and shares to the same could be linked to guaranteed access to power in proportion to the shares held by industry and others including the DCGs. This would only be an inferior solution, to outright purchase of IPPs, since the SPV would have a life of twenty years, and most importantly would also lock the system firmly to the cost of the inflexibility of fuel choice embedded in PPA contracts. It would also mean a major organizational effort on the part of the restructuring authority to create the SPV with an appropriate incentive structure for its owners and managers so that they could perform as anticipated.

8.4 THE MODALITIES

The success of the restructuring would crucially depend upon the process adopted and on fixing target dates for significant events in the process, much before the process begins.

The first step would be to appoint a disinvestment and restructuring commissioner (RC) with the necessary skills, and to whom the current proposal in its philosophy and details is acceptable. His terms of appointment would also need to be defined. He should be granted complete insulation from the political process, and from the day-to-day functioning of the government. The government would have to commit to adhere to the overall philosophy of the restructuring; to agree to pilot the required bills in the Assembly; to amend and repeal such acts as necessary and bringing in fresh legislation; to provide such resources as required for restructuring, and for subsidization, if so chosen.

The commissioner would have to complete his task in three years. He would have to have extraordinary powers to supersede the board of the GEB, and transfer top officials if necessary, so that the process of disinvestment is not hindered by vested interests. His office would need to be provided with funds to cover the cost of studies, preparation of legal documents, negotiations, consultant services, asset revaluation and modelling studies that may be required.

The commissioner would report once in six months to a board that would consist of parties representing the government, industrial consumers, and other power consumers. The board should also be able to modify the proposals of the commissioner only with two-thirds majority. This would be necessary to remove any doubts about back tracking, and to be able to implement the restructuring and privatization in a period of three years.

Even before the RC is set up, the government would have to take a decision to the effect that farmers be subsidized directly. This could be done via coupons distributed bi-annually through banks to all farmers who are issued identity cards, which also mention their eligibility levels. A special taskforce with officials from the agriculture, finance and home departments would have to be set up to bring about a fool-proof direct subsidization scheme. Consultants and contractors required to implement the scheme, and budgetary provisions to cover the costs of the same, should also be agreed upon.

Demand for power would, in the long term, grow at average rates that can be expected to be no less than 4.5 per cent, with a GDP growth of 5 per cent, which is the least that one can expect. In all likelihood, with the right macro-economic policies for growth, the power sector should grow at rates in excess of 6 per cent per annum for many years to come. Current surplus labour can be potentially absorbed within the system. Yet it would be desirable to quickly shed surplus labour to the extent seen fit by the inheritor companies. This would protect systemic efficiency and remove the distortions (in technology choice, managerial practices, and in assessment of costs). Politically though, a median path that allows a liberal VRS for all GEB staff, especially for office and commercial staff and in departments where the excess is large, would be appropriate. The government would have to announce upfront that the restructuring would not result in compulsory retrenchment. The actual exercise of implementing the VRS would have to be left to the inheritor companies, since they would be best able to assess the potential and value of employees. A fund to cover the VRS of as much as 25 per cent of the current staff would have to be created by the government, and made available to the inheritor companies.

The total wage, salary and terminal benefit costs of 25 per cent of the employees in 1996-7 (the latest year for which the annual report of the GEB was available) is Rs. 121 crore or Rs. 0.91 lakh per person for approximately 12,130 persons (excess manpower). Assuming a coverage of 50 per cent of current salary for the remaining work life of the employee, and at 50 per cent of the terminal salary for pension, and with a life expectancy of 72 years and retirement age of 58 years, the VRS cost (at 3 per cent real rate of interest) would be Rs. 1,119 crore, or Rs. 8.42 lakh per person. At 70 per cent coverage of salary it would amount to Rs. 1,567 crore and Rs. 11.8 lakh per person. Thus a fund of between Rs. 1,200 to 1,600 crore would be required to operationalize a near complete shedding of the excess manpower. A commitment to make available up to that amount for the VRS would have to be made by the government. An incentive of about one-tenth the amount should be allowed to the privatized companies that retain manpower in anticipation of future requirements, for a period of five years. A detailed study to assess the actual distribution of excess manpower across various locations and activities would be necessary. The second charge on the privatization proceeds would have to be towards coverage of VRS, the first charge being to 'buy' out the stranded contracts.

The government should announce its intention to privatize and restructure the GEB, along the lines in this document, even before the appointment of the Restructuring Commissioner (RC) or the Board for Privatization. The RC would then have the task of implementing the same. The actual details of bundling generating assets with distribution zones or territories, can be left to the bid-process, as long as the overall process and objectives of the reform are clarified early enough, and complete transparency is ensured, through appropriate processes. Thus the terms covering the appointment of the RC should bind him to transparent processes and to disclosure of information. All relevant information generated by the various studies, consultancy studies and reports, should be made public and available to the bidder. They in turn should be encouraged to make their own investigations and studies, especially with regard to excess manpower, state of assets and equipment, and extent and nature of demand. There is likely to be much asymmetry in value of assets and territories to potential bidders depending upon the nature of their complementary assets and skills.

8.5 CONCLUSION: STAGES IN THE RESTRUCTURING PROCESS

The following stages for reform are proposed:

— Approval of the white paper (this document) outlining the restructuring strategy of the government of Gujarat.
— Announcement of the overall strategy through the press and other media to the public, and potential bidders, and financial institutions.
— Discussions and consultations around the white paper.
— Appointment of a five-member Board for Electricity Restructuring and Privatization.
— Board meets to discuss and select the Restructuring Commissioner (RC); government approval sought for appointment of RC; RC takes over as chairman of the Board of Restructuring. RC submits his objections if any to this agenda, and a final agenda for restructuring and privatization is approved by the board.
— The task force for the development of a direct subsidization scheme set up by the RC in consultation with the government.
— Introduction of bills, and amendments in the Assembly; an issuance of appropriate orders to further the restructuring.
— Consultancies for studies, preparation of bid documents, cost studies and valuation, data generation, and laying out the privatization process. Board accords approval to detailed plans of the RC.
— Regulatory strategy announced by the GERC; including such details as power caps or ROI, the role of power imports from outside Gujarat over the long term. He would also have to announce the degree to which exchange risk, fuel price risk and such other costs are passthrough.
— The scheme of direct subsidization is in place based on power coupons and identity cards for farmers.
— Non-financial bids invited and the objectives of privatization publicized adequately. Bidders allowed near complete access to all available information including that generated by consultancy studies, and encouraged to carry out their own preliminary studies and assessments.
— Non-financial bids are examined and a shortlist of bidders selected for submission of financial bid, by the RC.

— TRANSCO is set up and its board and CEO selected by the RC. Other principal officers/ managers selected by its board. TRANSCO takes over the transmission assets of the GEB, and announces the details of transmission rules, connection charges, and penalties in consultation with the RC and GERC.
— IPP Contracts extinguished by retiring equity at 16 per cent over the life of the project. SPV to manage the financial inflows and outflows due to the restructuring set-up with the RC as chairman. SPV receives first installment of VRS funds from the government. SPV to manage (if IPP contracts are not extinguished) stranded contracts , or to recover the cost of extinction through appropriate notes to the GERC.
— VRS policy and strategy announced.
— Detailed and overlapping bids invited; financial assistance for preparation of bids provided to the shortlisted bidders.
— Clarifications, discussions and consultations take place, between the RC, the GERC, financial institutions and bidders.
— Detailed bids examined and the selected parties invited to take over generation and distribution assets of the GEB.
— Four DCGs, one HYGEN and several GCs (ex-IPPs and CPCs, CPs and Bulk Consumers register as part of the Gujarat Power Market, and sign appropriate contracts with the TRANSCO. Market-making processes for wholesale power implemented, further refinement and fine tuning of the rules.
— Proceeds of asset sale go into SPV.
— VRS comes into effect.
— Board of Restructuring and the RC wound up. The SPV (if need be) continues till the life of the stranded contracts are over and VRS is closed.

NOTE

1. Except to the extent of the distortions created by the IPPs with whom the state has signed long-term contracts favourable to the IPPs.

PART III

LEARNING ON 'MODELS'

9

Andhra Pradesh Power Sector Reforms: Issues of Transition

USHA RAMACHANDRA

Reforms were initiated in the Indian power sector in the early 1990s to mobilize investment in power generation to bridge the growing gap between power supply and demand.[1] However, it was soon realized that the sector was beset with financial, operational and organizational inadequacies, which needed to be addressed as part of the reform process. However, unless the transitional aspects of reforms are managed effectively, the reform process and its sole motivation of mobilizing private investment would fail. Handling the transition is therefore very important both for the success of reforms as well as for creating an environment conducive to reforms.

This chapter first lays down the context and contours of power sector reforms in the state of Andhra Pradesh. It then outlines the several transitional issues in the context of what was planned as well as what actually took place. Finally, the chapter attempts to examine whether these problems could be overcome within the framework of reforms as initiated in Andhra Pradesh or whether the reform process itself requires to be re-examined.

9.1 BACKGROUND

In the early years of the electricity sector before Independence the power industry in Andhra Pradesh was—as was the case in several other states across the country—predominantly urban-based and in the private sector. With planned economic development initiated at the time of independence it was necessary to respond to the power needs of the rural areas through public investment and public

TABLE 9.1: INSTALLED CAPACITY—ANDHRA PRADESH

(year: 2000)

Sl. No.	Source	Type	Capacity (MW)
(a)	State Sector (AP GENCO)	Thermal	2952.50
		Hydel	2672.00
		Wind	2.00
	Total State Sector		*5626.50*
(b)	Joint sector	Gas	272.50
(c)	Private sector	Gas	424.00
		Mini Hydel	43.45
		Wind, Cogen & others	193.05
	Total Private Sector		*660.50*
(d)	Central sector (share)	Thermal & Nuclear	942.50
	Total (a+b+c+d)		7502.00

Note: In addition, Andhra Pradesh has captive generating capacity of about 1,500 MW mainly with major and medium industries and large commercial consumers.

ownership. Towards this end, the government enacted a new law under which the power sector would be owned, managed, regulated and operated by the government, i.e. the Electricity Supply Act 1948 (ES Act).[2]

The rapid development of the power sector in Andhra Pradesh started with the formation of the Andhra Pradesh State Electricity Board (APSEB) in April 1959 as a statutory body under the provisions of the ES Act (1948). APSEB was a vertically integrated utility with the responsibility for electricity generation, transmission and distribution (T&D) for the entire state of Andhra Pradesh and was one of the larger utilities in India. In the year 2000, APSEB had an total installed capacity of 5,627 MW, T&D lines of 6,37,885 km, more than 1 crore consumers and a total sales of 27,605 MU. By January 1999, APSEB was a very large utility with assets over Rs. 12,300 crore (see Table 9.1).

There was unrestricted availability of power till 1978 and until 1990, APSEB was one of the few SEBs in India which had a continuous record of profitable operations, though the profits came up to the mandatory 3 per cent cf fixed assets only in a few years. However, the financial position of APSEB deteriorated rapidly from 1990 onwards and the state government was forced to provide subvention year after year in order to enable APSEB get a 3 per cent return on fixed assets.[3] The major reason for this crisis was

TABLE 9.2: PERFORMANCE OF APSEB/TRANSCO

(Rs. in crore)

Year	Energy Sold (mu)	Total Revenue	Expendi-ture	Profit/ Loss (-) w/o subsidy	Subsidy	Profit/ Loss (-) with subsidy
1980-1	5104	204.57	216.28	(11.71)	11.71	-
1985-6	10312	568.86	578.51	(9.65)	24.01	14.36
1990-1	16093	1262.94	1220.71	42.23	38.76	80.99
1994-5	23096	2276.33	3133.19	(856.86)	944.11	87.25
1997-8	23944	4363.41	5497.80	(1134.39)	1255.94	121.55
1998-9	25232	4621.95	6997.84	(2375.89)	2504.53	128.64
TRANSCO						
1999-2000*	25800	5000.00	8000.00	(3000.00)	-	(3000.00)
2000-1*	26500	6000.00	8926.00	(2926.00)	1626.00	1300.00

Source: AP TRANSCO
* Provisional.

TABLE 9.3: EXPENDITURE – REVENUE – SUBSIDY – TARIFF REVISION

(Rs. in crore)

Description	1994-5	1995-6	1996-7	1997-8	1998-9	1999-2000 (Prov.)
Expenditure	3133	3572	4280	5498	7036	7375
Revenue Recovered	2276	2443	3558	4363	4629	4875
Revenue Deficit/Loss	857	1129	722	1135	2407	2500
3% ROR on Net Fixed Assets	87	130	129	122	112	172
Total Subsidy	944	1259	851	1256	2519	2672
Subsidy Paid/adjusted by GoAP	Equity written off	Adjusted against Loan	Adjusted against Loan	Rs. 483 crore Adjusted against Loan	Cash Loan of Rs. 545 crore Bonds Rs.1100 crore Total Rs.1645 crore	Cash grant of Rs.105 crore + Bonds for Rs. 3013.24 crore raised
Tariff Revision (Date)	-	19% (2.8.95)	32% (1.8.96)	-	10% (31.12.98)	-

Source: Power Sector Status and Tariff in Andhra Pradesh (2000-1), Andra Pradesh Electricity Regulatory Commission (APERC).

the politicization of most decisions of the Board as well as a distinct lack of commercial orientation. Due to frequent elections and political uncertainty and the significant role that electricity prices play in vote-bank politics, tariffs were not revised for several years. Added to this was the rapid increase in T&D losses on account of increasing indiscipline and lack of accountability on the part of the field staff and large-scale pilferage. The net effect was a rapid deterioration in the finances of the APSEB, as illustrated in Tables 9.2 and 9.3.

Commercial T&D losses (a euphemism for theft) was camouflaged as agricultural consumption which was not metered and hence could not be verified. And total energy billed was continuously in the range of 39-41 per cent (see Table 9.4).

The figures of agricultural consumption were adjusted to increase at a very rapid rate to cover up the increasing losses on T&D. After a detailed study and survey AP TRANSCO itself corrected the figures 1996-7 and brought down the specific consumption per pumpset as per Table 9.5.

TABLE 9.4: PERCENTAGE OF UNITS BILLED TO THE TOTAL ENERGY PURCHASED

S. No	Description	1994-5	1995-6	1996-7	1997-8	1998-9	1999-2000
1	Domestic	11%	11%	12%	12%	13%	14%
2	Non-Domestic	2%	2%	2%	3%	3%	3%
3	Industrial	4%	4%	4%	4%	4%	4%
4	Public Lighting	1%	1%	1%	1%	1%	1%
5	High Tension	23%	22%	22%	20%	18%	18%
	Total	41%	41%	41%	40%	39%	41%

Source: Power Sector Status and Tariff in Andhra Pradesh (2000-1), APERC.

TABLE 9.5: CONSUMPTION FOR IRRIGATION

Sl. No	Item	1990	1995	1996	1997	1998	1999
1	No. of pumpsets (lakh)	10.57	15.30	15.63	17.06	17.36	17.92
2	Total energy consumption in Agricultural (MU).	5276	10922	11399	7835	9336	9866
3.	Consumption per pumpset (kWh)	4994	7138	7592	4592	5377	5505

Source: Power Sector Status and Tariff in Andhra Pradesh (2000-1), APERC.

Consequently the T&D loss figures had to be corrected upwards which resulted in an erroneous public perception that the T&D losses increased with the stepping up of reforms (Table 9.6).

At the same time, given the lack of autonomy to fix tariffs by SEBs, the political compulsion to subsidize rates for agriculture and domestic consumers led to extremely skewed tariffs. The average tariffs for various categories of consumers in 1976-7 are given in Table 9.7.

Firstly, the share of agriculture in total consumption rose from 24.91 per cent in 1985-6 to 39.24 per cent in 1998-9 (even after 'correcting' for T&D losses) a growth of almost 11 per cent while that of industry grew by only 2.75 per cent. This was not due to a

TABLE 9.6: T&D LOSSES

(in MU)

Year	Power Generated	Power Purchased	Total Energy Handled	Power Sales	Loss of Power	Loss (%)
1994-5	19,500	9,129	28,629	23,095	5,534	19.33
1995-6	20,417	8,732	29,149	23,562	5,587	19.17
1996-7	23,099	8,501	31,600	21,068	10,532	33.33
1997-8	24,631	11,187	35,818	23,944	11,874	33.15
1998-9	25,246	12,366	37,612	25,224	12,388	32.94
1999-2000	27,394	13,365	40,759	27,523	13,236	32.47
2000-1 (Projected)	27,221	15,407	42,628	27,538	15,090	35.40

Source: Power Sector Status and Tariff in Andhra Pradesh (2000-01), APERC.

TABLE 9.7: AVERAGE TARIFF: ANDHRA PRADESH—1996-97

(Ps/kWh)

Consumer Category	Tariff
Domestic	142.50
Commercial	327.66
Agriculture/Irrigation	13.35
Industrial	309.24
Railway Traction	329.23
Outside State Sales	87.24
Overall Average	156.00

Source: Annual Report on the Working of State Electricity Boards and Electricity Departments, Planning Commission, GOI, June 2001.

TABLE 9.8: ANDHRA PRADESH: CATEGORY-WISE CONSUMPTION

Sl. No.	Item/Category	Residential	Commercial	Agriculture	Industry	Others	Total
1	Consumption (MU)						
	1985-6	1195	340	2569	6089	119	10312
	1998-9	5090	1060	9866	8655	472	25143
2	Share of Consumption (%) in total						
	1985-6	11.59	3.30	24.91	59.05	1.15	100.00
	1998-9	20.24	4.22	39.24	34.42	1.88	100.00
3	Average Annual Growth 1985-6 to 1998-9						
	(a) Energy (MU)	300	55	562	197	27	1141
	(b) Growth (%)	11.80	9.10	10.90	2.75	11.20	7.10

Source: Power Sector Status and Tariff in Andhra Pradesh (2000-1), APERC.

TABLE 9.9: CONSUMPTION OF HT INDUSTRIES INCLUDING CAPTIVE GENERATION (1989-9)

(in MU)

Sl. No.	Item	1989	1995	1996	1997	1998	1999
1	SEB Sales to HT Industrial and Non-Industrial (%)	5477 (77.01)	6763 (69.27)	6604 (63.81)	6914 (63.60)	7216 (65.66)	7136 (62.19)
2	Captive Generation by Industries (%)	1628 (22.99)	3000 (30.73)	3745 (36.19)	3957 (36.4)	3773 (34.34)	4339 (37.81)
3	Total Electricity Consumption by HT Industrial and Non-Industrial	7105	9763	10349	10871	10989	11475

Note: Percentage figures have been rounded off.

decline in industrial consumption per se but attributable to an increase in captive generation by industrial consumers (see Tables 9.8 and 9.9).

9.1.1 Chronology of Reforms

Table 9.10 shows the main steps of reforms. Alarmed at the fast deteriorating finances of the APSEB, the state government set up a High Level Committee under the chairmanship of Hiten Bhaya in early 1995 to advise and provide guidelines for the restructuring of

TABLE 9.10: CHRONOLOGY OF EVENTS

April 1959	Formation of Andhra Pradesh State Electricity Board.
January 1995	GoAP sets up Hiten Bhaya Committee to advise and provide policy guidelines for restructuring the AP electricity industry. The Committee submits its reports three months later.
December 1996	Chief Ministers conference and the announcement of the *Common Minimum National Action Plan for Power.*
February 1997	GoAP Policy on Power Sector Reforms.
January 1999	AP Electricity Reform Transfer Scheme and Rules notified.
February 1999	Andhra Pradesh Electricity Reform Bill enacted.
February 1999	Andhra Pradesh Electricity Regulatory Commission set up.
March 2000	AP TRANSCO restructured into a holding company and four distribution companies.

the electricity sector in the state. In April 1995, the committee submitted its report that suggested the unbundling of the APSEB into a generating company, a transmission company and several (4 or 5) distribution companies. It also proposed an independent regulatory commission. There appear several reasons for suggesting such radical restructuring. The significant reasons were that the functional priorities get distorted in a vertically integrated utility, motivation and morale in the monopoly utility was low, there was a lack of responsiveness to consumer problems, and interference by the government had eroded the commercial orientation of the Board which in fact was mandated under the ES Act 1948.

The Government of Andhra Pradesh (GoAP) published a White Paper on the finances of APSEB and sought public opinion. Meanwhile, the imperative need for power sector reforms were recognized by a national consensus and a 'Common Minimum National Action Plan for Power' were announced after a meeting of chief ministers of states was held on 3 December 1996 at New Delhi. Several other states predominantly, Orissa, Haryana, Karnataka and Rajasthan voiced the need for power sector reforms in their respective areas.

The GoAP formulated its power sector reform policy and announced it in February 1997. The main components of the proposed reform programme were:[4]

— Segregation of the generation and transmission and distribution functions into separate autonomous companies.

— Reconfiguration of the distribution system into distinct

geographical zones each served by an autonomous distribution utility.

- Corporatization and commercialization of the emerging power sector entities, by making it attractive for Indian and foreign investors to readily invest in the power sector; to establish the needed capacities in generation, transmission and distribution; to meet the growth needs.
- Establishment of an independent regulatory commission to regulate the functioning of the power sector on sound commercial principles; to safeguard the interests of the customers in respect of quality, reliability and fair price for electricity; to set cost and efficiency-based tariffs to ensure the creditworthiness and viability of the power sector and to progressively eliminate tariff distortions and subsidies.
- Retaining with the state government the role of policy formulation.
- Progressively promoting competition among various power sector entities.

For achieving these objectives, the GoAP announced its intention to restructure the APSEB on functional lines into companies dealing with generation, transmission and distribution of power. To ensure fair play and equity between the companies and the consumers in an unbundled power sector, an independent regulatory commission was proposed to be set up under an appropriate statute.

At about the same time, the GoAP decided to seek the technical and financial assistance of the World Bank to initiate economic reforms in the state as a whole, and its plans for the electricity sector became part of this agenda. The World Bank arranged through several donor-funding agencies, to provide a matrix of technical assistance support for power sector reforms. Broadly, the technical assistance support was as follows and the amount of grants exceeded over US $200 million (see Table 9.11).

The World Bank's assistance to the state's power sector reforms through the Adaptive Lending Procedure (APL) exceeded a total of US $1 billion. The conditionalities for each tranche of loan was clearly laid down in the Project Appraisal Document of the World Bank and provides a clue to why reforms took the path they did in Andhra Pradesh.[5]

The policy on 'Reform in the Power Sector in Andhra Pradesh' set up a three-tier Reform Implementation Organization:

TABLE 9.11: TECHNICAL ASSISTANCE SUPPORT FOR POWER SECTOR REFORMS

Assignment	Funded by	Terms
Institutional Strengthening and Human Resources Development	DFID	Grant
Corporatization and Commercialization	DFID	Grant
Developing Privatization Strategy, selection of Joint Ventures partners and Assisting GoAP in fund raising	DFID	Grant
Financial Restructuring	WB	Loan
Load Forecast, Tariff Studies DSM & Technical Interface	CIDA	Grant
Consultants to APERC	WB	Loan

Note: Most grants were towards studies and capacity building though a small portion was also for distribution augmentation.

1. A steering committee of senior officials concerned and other experts presided over by the chief secretary.
2. Task force of officials of the SEB and Department of Energy of the Government presided over by the energy secretary, and
3. Working groups of SEB officials supported by retired senior employees (Chief Engineers)

Though the GoAP was ready to commence work in 1997, the entrustment of the studies to the consultants led to a delay in the implementation of reform measures, as several groups were providing advice on diverse issues and not necessarily in a cohesive manner. However, as a result of the efforts of the consultants and of the working groups of SEB officials, a comprehensive Electricity Reform Bill was formulated and passed in April 1998 by the legislature. But the Act was not notified for a number of months, as there was some delay in getting the assent of the President of India. It was notified on 27 January 1999 to come into effect from 1 February 1999.

On 1 February 1999 the Andhra Pradesh Electricity Regulatory Commission (APERC) was set up and the APSEB was unbundled into AP GENCO (A.P. Generation Company) and AP TRANSCO (A.P. Transmission Company) with an indication that distribution would be soon hived off into several smaller companies.

The APERC took over several of the regulatory functions which had earlier rested with the GoAP and the APSEB. Its powers included issuing licenses to the transmission and distribution companies, regulating tariffs, promoting competition, fixing standards of performance and protecting the interests of the stakeholders.[6] The GoAP retained its powers to formulate overall policy guidelines, plan for capacity creation and provide subsidies in a transparent manner through the budgetary process where required.

Under the new regulatory regime, AP TRANSCO filed for a tariff revision before the APERC over December 1999 and January 2000. Even while the process of tariff setting was taking place, on 22 March 2000, the GoAP announced the restructuring of AP TRANSCO into a transmission corporation with four wholly-owned subsidiary distribution companies. This decision led to the speculation, not entirely unjustified, that the reform process in the power sector was more in line with adhering to the milestones of the WB loan agreement rather than a rational process of introducing reforms.

9.2 ISSUES IN THE TRANSITION

The performance of the AP TRANSCO—the successor organization to APSEB—has deteriorated even while the process of reforms are underway. This paper will attempt to analyse the problems of transition along the following issues:

1. The reforms process
2. Stakeholders' participation
3. Organizational structure
4. The new regulatory framework
5. Dealing with the past

9.2.1 The Reforms Process

The Andhra Pradesh Electricity Reform Act passed in 1998 was substantially the same as the Orissa Electricity Reform Act, which was passed in 1995. Andhra Pradesh first initiated discussion on power sector reforms in 1995 but took more than two years to provide the basic legal foundation for the reform. The delay could have and should have been avoided, as during this period, the employees, uncertain about their future, became increasingly demoralized and the trade unions were active in propounding theories of GoAP and APSEB totally 'surrendering to World Bank which was directing the course of reforms'.

As seen above, from 1995 to 1999 there were very intensive consultations but the consumers were in the dark. Some glossy brochures were brought out but their circulation was confined to the elite. In any case, these brochures did not contain material that could 'sell' reforms to the consumer. On 1 February 1999, a number of changes were brought about on a single day. The Andhra Pradesh Electricity Reform Act, 1998 was brought into effect. On the same day, the Andhra Pradesh Electricity Regulatory Commission (APERC) was set up. The Andhra Pradesh Electricity Reform Transfer Scheme and Rules had been notified on 29 January 1999; came into effect from 1 February 1999. Under this the Andhra Pradesh State Electricity Board was converted into Andhra Pradesh Generation Corporation (AP GENCO), and the Transmission Corporation of Andhra Pradesh (AP TRANSCO), with the indication that as soon as possible AP TRANSCO would hive off distribution into several small companies.

All these changes were brought about through the Andhra Pradesh Reform Transfers (Transfer Scheme Rules, 1999). These rules were issued by the government under Sections 23, 24 and 25 of the Andhra Pradesh Electricity Reform Act of 1998. This Act has a separate part (Part V) dealing with the restructuring of the electricity industry. Section 13 makes it mandatory that the state government constitutes within sixty days of the Act, AP TRANSCO as the transmission corporation with the principal objects of engaging in the business of procurement, transmission and supply of electric energy. Under Section 23, the state government has to prepare a transfer scheme under which, from the date notified, all the assets of the APSEB would be vested in the state government. Any property, once vested in the government, can be re-vested in either Andhra Pradesh TRANSCO or in any Generating Company or Companies in accordance with this scheme. A transfer scheme should provide for formation of subsidiaries or joint venture companies and it can define the property rights and liabilities to be allocated. The Reform Act provided sixty days time between the date of bringing the Act into effect and the date of effecting the transfer scheme. This time could have been taken advantage of, to have wide consultations on the transfer scheme. The consumers could have brought up the arguments for having a separate hydroelectric and thermal generation corporation and even separating the transmission and distribution functions on the same date as suggested by the High Level Committee. Whatever was the final decision, consumer participation in these decisions would have prepared them for the subsequent stages of power sector reforms. But this was not done.

A tripartite agreement was entered into with the various unions of APSEB workers, the GoAP and the SEB to regulate the allocation of the employees in restructured organizations. The transfer scheme (AP Electricity Reform-Transfer Scheme Rules, January 1999), which created the unbundled companies, also set out the principles under which the assets of APSEB were divided between AP GENCO and AP TRANSCO. The first transfer of assets was done on the basis of valuation at book value. All the assets, which were directly related to the function of generation, were separated and brought under AP GENCO. The AP GENCO got all the power stations which comprised 2657 MW of hydel and 2972 MW of thermal power stations. All the other assets were vested under AP TRANSCO. The government retained about Rs. 3812 million of liabilities and divided the remaining liabilities to AP TRANSCO and AP GENCO in proportion to their assets. Broadly the assets and liabilities inherited by the two companies under the transfer scheme are illustrated in Table 9.12.

Though the value of assets distributed was set out in the Transfer Scheme, it was indicated that these were subject to revision within one year! The transfer of personnel was done as per the conditions and terms contained in the Act and the tripartite agreement, which

TABLE 9.12: DIVISION OF ASSETS AND LIABILITIES BETWEEN AP GENCO AND AP TRANSCO AS PER TRANSFER SCHEME, JANUARY 1999

(Rs. in crore)

	GENCO	TRANSCO
Net Fixed Assets	2719.7	2586.2
Capital Works in Progress	2380.1	705.9
Investments	—	44.1
Total Fixed Assets	5099.8	3336.2
Current Assets	1506.4	2073.7
Subsidy Receivable from Govt.	—	772.9
Total Assets	6606.2	6182.9
Equity Capital	780.2	540.9
Grants for Capital/Consumer Contribution	71.5	443.9
Capital Reserve	23.3	4.8
Long Term Loans	3583.7	2962.2
Current Liabilities	2211.5	2231.1
Total Liabilities	6606.2	6182.9

was entered into with eight employee unions. The personnel under this agreement have been given five years time during which they could opt for a change from one company to another!

The pension liability of the erstwhile APSEB employees became a bone of contention between the union and the restructured corporation. Private sector corporations make such provisions annually in the accounts and show it as a 'fund'. APSEB had not done so and the pension payable was met yearly as part of the wage bill of employees. This is the procedure in the government and the APSEB followed government procedures. In view of the legal requirement that corporate bodies acknowledge the pension liability, the APER Transfer Scheme includes a provision for the government making appropriate arrangements to fund the pension fund and other personnel related funds separately, as these liabilities were not specifically provided in the schedule, which listed the transferred assets and liabilities. The scheme also allows a period of one year up to which the terms and conditions and clarification of the assets would be treated as provisional and could be changed by the government. At the end of one year from the date of transfer, the clarification and terms of transfer become final. The 'terms' includes 'value' of assets. The period of one year of 'provisional' terms of transfer expired on 1 February 2000.

On 31 January 2000, just a day before the expiry of the one-year period, GoAP issued the 'Amendments to APER (Transfer scheme), Rules 1999'. Under this, the value of gross fixed assets of AP GENCO was increased by a hefty Rs. 4,705.4 crore.[8] There is a countervailing entry on the liabilities side of Rs. 4,386.9 crore shown as 'Pension Fund' (see Table 9.13).* Obviously 'the pension fund and personnel related funds', have been estimated by actuarial calculations to amount to Rs. 4,386.9 crore, the government has attached that liability to AP GENCO and revalued its assets upwards especially the hydropower generating plants. Such a revaluation has a serious impact on consumer tariff and the consumers are likely to challenge it as no justification or basis for revaluation has been given in the notification of the Amendments to the Transfer Scheme.

*This is seen from compairing Tables 9.12 and 9.13. Indeed net fixed assets jump from Rs. 2,719.7 crore to Rs. 4,705.4 crore. Similarly, total liabilities increase by Rs. 4,559.4 crore of which Rs. 4,386.9 crore relate to pension liabilities.

TABLE 9.13: DIVISION OF ASSETS AND LIABILITIES BETWEEN AP GENCO AND AP TRANSCO AS PER TRANSFER SCHEME 1999 (REVALUED ON 31-2-2000)

(Rs. in crore)

	GENCO	TRANSCO
Net Fixed Assets	7425.1	2654.6
Capital Works in Progress	2250.1	841.2
Investments	—	37.5
Total Fixed Assets	9681.8	3533.3
Current Assets	1483.8	1570.2
Subsidy Receivable from Govt.	—	1202.4
Total Assets	11165.6	6245.9
Equity Capital	2106.8	540.6
Grants for Capital/Consumer Contribution	71.5	765.8
Capital Reserve	23.3	4.8
Long Term Loans	6860.8	1784.9
Current Liabilities	2198.6	3920.4
Total Liabilities	11165.6	6245.9

No doubt, what has been done is one way to accommodate the pension fund liability. It was, however, not the best method, especially from the consumer's perspective. Nor was it the most opportune moment to revalue assets to almost twice its original value without a full and proper justification for the revaluation being given in the government orders.

9.2.2 Stakeholders' Participation

One of the most important component of any reforms process is continuous dialogue with the public. However, the process of providing information to the public through the media and personal workshops became neither the responsibility of the GoAP, the APERC or the utilities themselves.

AP TRANSCO—Consumer Interface: The purpose of reform in the power sector was to give a better deal to the consumers. It is very strange that on the day of the formation of AP TRANSCO, and the abolition of the APSEB, there was not a single advertisement in the newspapers about AP TRANSCO and how it would be different in its dealings with the customers. Over the years, the customers have signed agreements with the APSEB and on the formation of AP TRANSCO, a letter could have been sent by the new company that the agreement executed by the customers now stands transferred to

AP TRANSCO and that the new company would attempt to give better service to the consumers. Even with the first bill that was sent out to the customers from AP TRANSCO, except for the change in the title of the bill, there was no indication that a private company had taken over the functions of the erstwhile APSEB. A great opportunity to win consumer support was thus lost.

Even before the formation of AP TRANSCO, there was talk of the APSEB publishing a Citizen's Charter. AP TRANSCO, however, did publish a booklet on 'Service Standards' of AP TRANSCO in 2000. It has set out ambitious levels of efficiency such as rectification of streetlights in 24 hours, replacing of fused bulbs in 7 days, replacement of failed transformers in cities in 24 hours, burnt meters to be replaced within 7 days, etc. It appears that neither AP TRANSCO nor the customers have taken serious note of these 'standards', as there is no indication of any penalty on AP TRANSCO for default.

The only change that customers have observed, is that. AP TRANSCO has launched a very intensive campaign against theft of electricity; schemes under which rewards are given to those who provide information about theft of power have also been announced. The response so far has been fair. The initiatives taken to reduce theft of electricity by other means have also yielded some positive results. However, there has been no discussion on the issue of defaulting of bills, probably because the government (and allied organizations) would have topped the list.

AP TRANSCO published a white paper in December 1999 which it said would be its communication to consumers, and which it intended to bring out every month. The paper gave details about average consumption and how many transformers were erected, etc. It did not give information about the interruptions in service, the number of complaints registered, about low voltage, etc. This could have been a good beginning for a monthly newsletter to consumers. It could have provided information to the general public through the electronic media and the newspapers. However, the first issue was also the last.

Government-Utility Interface: With the separation of transmission and distribution and corporatization as AP TRANSCO it was hoped that the company would distance the operations of electricity supply from the government and come closer to the consumers. Unfortunately the composition of the boards of the new company gives ample indication that such a hope would never materialize. The chairmen of the two companies were the former chairman and member secretary of the APSEB, and the former Members of the APSEB were divided in the two companies as directors, with only the energy secretary, GOAP as the additional member on both the boards. Even after the distribution companies were formed there has been no move to constitute a full-fledged board with management expertise. This made sure that there was no likelihood of any fresh approach to commercialization and consumer relations.

Throughout the period from February 1999 to March 2000, there were reports in the press of the chief minister himself monitoring the operations of the new companies almost every day. In fact several announcements, by ministers of GoAP were on matters which were not policy-related but mostly operations-related. Early in January, 2000, answering questions about the possible increase in tariff, one minister indicated that tariff would be increased by 15 per cent without even mentioning the fact that the APERC has already been set up and all issues relating to tariff would be decided by the Commission. It was obvious to the consumers that the GoAP had no intention to distance itself from AP TRANSCO and its subsidiaries.

The issues of separating distribution companies from transmission companies were studied by a large number of consultants, but their work was known only to a handful of AP TRANSCO employees who were in the reform cell of AP TRANSCO; thousands of engineers and employees never had an opportunity to discuss any reform issue openly. Needless to say the consumers were only given occasional press statements and booklets. Considering that consumers were not made active participants in the discussion of the reform proposals, the acrimonious debates on minor details of the operations raised by political parties was primarily on account of the perception that the chief minister was taking all the decisions including those that related to day-to-day operations.

However it is to the credit of AP TRANSCO that it maintained the supply of electricity to the consumers at usual if not at slightly better levels in spite of a steep increase in demand particularly from the agriculture sector, coupled with a problem of diminished supply. AP TRANSCO achieved this by importing power from states which had a saleable surplus, at higher prices. Senior management of AP TRANSCO was fully tied down to the task of arranging for import of power everyday and raising money to meet the power purchase bills. AP TRANSCO succeeded in both these efforts at the cost of financial prudence. Furthermore, these chores left very little time to AP TRANSCO to improve its customer relations.

AP TRANSCO-Regulator Interface: Both AP TRANSCO, which came into existence on 1 February 1999 and the APERC, which became functional from 3 April 1999, are young organizations and the issues of interface of AP TRANSCO with the regulatory commission are slowly surfacing.

AP TRANSCO has dealings with the APERC in matters of:

1. Obtaining a supply license (Sec. 15 of APR Act, 1998)
2. Approval of the required revenue and retail tariff (Sec. 11(e) of APR Act)
3. Approval of PPAs (Sec. 11(e) of APR Act)
4. Approval of applications for captive power generation by industrial consumers (Sec. 21(3) of APR Act)
5. Approval of the standards of service, which APERC has to follow.

The licenses for AP TRANSCO have been issued already by the APERC for transmission and supply. AP TRANSCO had applied in the prescribed form and after hearing objections, APERC has issued separately one 'Transmission and Bulk Supply License' and another Distribution and Retail Supply License. Both these licenses were issued on 31 January 2000 and are valid for thirty years from 1 February 2000.

As APERC has powers to regulate the purchase of power, its supply and distribution, APERC has to approve any purchase agreement between AP TRANSCO and any generating company including IPPs and AP Genco.

The basis for approving any new PPA would be the electricity load forecast and resource plan which identifies the particular location, site and size of the proposed plant. Under the orders recently notified by APERC Guidelines for Load Forecasts, Resource Plans and Power Procurement dated 29 February 2000, any long term purchase of power either from a generating company within the state or from any other source has to be done with the permission of the APERC for which an application has to be made sixty days in advance. This however does not affect AP TRANSCO making short term purchases for emergency support and to realize short term operating cost reductions. The long term purchases would be approved only on demonstrating that it is being done in an economical, efficient and transparent manner through a structured competitive procurement process prescribed by the Commission. The standard structured competitive procurement process indicated by the APERC is on the same lines as the procedures of the World Bank except that in place of international competitive bidding, a national bidding process was suggested.

A lot of inconvenience and some amount of embarrassment would be caused by these orders to AP TRANSCO as a large number of PPAs are pending final approval and many IPPs are proposing changes in tune with changes in capital costs and perception of risk. As these have not been approved finally by 3 April 1999, when the Commission came into existence, they would have to come up to the Commission before approval by the government. The Commission has to call for public objections to these projects. Similarly where PPAs have been signed but financial closure has not been effected before 31 December 1999 would have to be de novo renegotiated.

The ability of AP TRANSCO to meet the growing power demand from private producers and the financial implications of such purchases is a matter of much speculation. The content of PPAs signed or initialled by the APSEB so far (which have been transferred to AP TRANSCO) were till recently confidential. However, when M/s BPL Power Projects sought an amendment in its PPA, the APERC had a public hearing and brought out an order requesting AP TRANSCO and M/s BPL to renegotiate on several counts before approaching the commission for approval. The issue really was that AP TRANSCO had no freedom to negotiate contracts on behalf of consumers without a public hearing.

Regulator-Consumer Interface: The APERC kept a low key in the first two years of its existence. The perception was that it would be like a court where cases would be held and decided rather than a regulator balancing the interests of all the stakeholders where information dissemination would be a significant part of its duties. The APERC is perceived as a government body rather than an independent commission on account of its initial decision to ban captive generation to protect the interests of the regulated, government-owned AP TRANSCO.

The most significant activity of any Commission is the tariff setting exercise that is conducted with stakeholder participation. However, the tariff decision of the APERC created an uproar in the state for two main reasons. First, APERC did not allow the media to be present during the 'public hearings' ostensibly on account of lack of space leading to allegations that the entire process was an eyewash and second, its controversial decision on tariffs for domestic consumers. However later public hearings have been held in large auditoriums with ample space for the public and the press.

9.2.3 Organizational Structure

As in most states that have restructured the power sector, Andhra Pradesh too unbundled into separate functional entities. The generation part of the electricity business was formed into AP GENCO and initially, transmission and distribution remained with AP TRANSCO. Later in March 2000, four subsidiary distribution companies were formed under AP TRANSCO. Andhra Pradesh, as in most other states that have unbundled, has adopted the single buyer model, wherein AP TRANSCO, the GoAP owned transmission company buys all the power from generating sources, pools the cost and sells it at a bulk tariff rate determined by the APERC.

The single buyer model is extremely popular in developing countries initiating power sector reforms. As transmission and dispatch are strategically very important functions, they continue to be with the government through the government-owned TRANSCO. The liabilities of power purchase agreements (PPAs) with generators are vested in this company and thus shield financiers of generation projects from market risk. And most importantly it helps maintain a unified retail price for electricity, which is mandated under the A.P. Electricity Reforms Act 1998. While the advantages of the single buyer model were highlighted when reforms were initiated, the disadvantages are now being perceived. With the government continuing to hold a powerful grip on the sector, it is often reluctant to impost financial discipline on distributors which in turn leads to low payment discipline. And powerful interests can indefinitely delay the next step towards the creation of markets for electricity. In Andhra Pradesh, the single buyer model has enabled the government to create an illusion of reforms without any substantive changes in the interim period. So far there has been no indication that the present single buyer model is a transition model and reforms would eventually lead to a more competitive market structure.

GoAP did not issue any official statement on the progress of the work of forming smaller distribution companies even while the local newspapers kept reporting odd pieces of information. In early January 2000, it was reported that the consultants to AP TRANSCO who had worked for over a year trying out a large number of alternative distribution models with the objective of bringing the average revenue realization of the distribution companies to absolute equal levels made their proposal to the GoAP. History and politics which binds certain districts together and administrative convenience and consumer service needs were not given adequate consideration. The proposals that were suggested by the consultants on the formation of the four distribution companies involved each company having a few districts which were not contiguous; in fact, some districts in the southern corner of the state belonging to a region called Rayalaseema were proposed to be linked to certain districts in the north which were a part of the coastal Andhra region. It is learnt that these proposals did not meet with the approval of the Government and the consultants were asked to review other alternatives with certain design criteria such as contiguity of the area under each distribution company.

Suddenly on 22 March 2000, the chief minister of AP announced in the state legislature the decision to restructure AP TRANSCO into a holding company with the same name and with four wholly-owned subsidiaries dealing with the distribution and supply functions. This announcement was followed on 27 March 2000, by a government order Ms. No. 31, from the Energy (Power III) Department, GoAP (Table 9.14).

The timing of the formation of the discoms was not quite appropriate. Under the new regulatory regime, AP TRANSCO filed for a tariff revision before the APERC over December 1999 and

TABLE 9.14: THE NEW DISCOMS

Name of Discom	Composition of Districts	HQ
Discom 1–NPDCL	Srikakulam, Vizianagaram, Vishakapatnam, East Godavari & West Godavari	Vishakapatnam
Discom 2–SPDCL	Krishna, Guntur, Nellore, Prakasam, Chittor, Cuddapah	Guntur
Discom 3–CPDCL	Kurnool, Anantapur, Hyderabad, Rangareddy, Nalgonda, Medak, Mehaboobnagar	Hyderabad
Discom 4–WPDCL	Adilabad, Nizamabad, Warangal, Karimnagar, Khammam	Warangal

January 2000. Even while the process of tariff setting was taking place, on 22 March 2000, the GoAP made the announcement of forming four distribution companies wholly owned by AP TRANSCO, the holding company. This announcement led to speculation, not entirely unjustified, that the reform process in the power sector was more in line with the conditionalities and milestones of the World Bank loan agreement than a rational process of introducing reforms.

9.2.4 The New Regulatory Framework

On 1 February 1999 the Andhra Pradesh Electricity Regulatory Commission (APERC) was set up, on the same day that the APSEB was restructured. The APERC took over several of the regulatory functions, which had earlier rested with the GoAP and the APSEB. Its powers included issuing licenses to the transmission and distribution companies, regulating tariffs, promoting competition, fixing standards of performance and protecting the interests of the stakeholders.

The APERC's first significant decision was in regard to the captive or self-generation policy. Keeping in view the precarious financial position of AP TRANSCO and that it could ill-afford to lose its 'paying' industrial consumers, the Commission decided to ban all future self-generation except for continuous process and frequency sensitive industries. In other institutions like hospitals where continuous power supply is essential, captive power would be allowed. But all captive generation would be for standby purposes only.[10] This decision of the Commission created uproar amongst the industrial community in the state and was challenged in court even while there were allegations of a regulatory capture. APERC is presently the only commission in the country, which is involved in the licensing of captive generation.

APERC as with all ERCs in the country has to deal with the best way of rationalizing tariffs even while avoiding a rate shock to any category of consumers. The revenue gap of the regulated utility is bridged by a combination of additional revenue from revised tariffs, subsidy from the state government and efficiency improvements of the utility especially in terms of reduction of T&D losses. In the first tariff hearing of APERC for TRANSCO for the year 2000-1 the GoAP committed to a subsidy of Rs. 1345 crore (see Table 9.15).

TABLE 9.15 CATEGORY-WISE LEVELS OF SUBSIDY

(*Rs. in crore*)

Name of the Category	Subsidy allocated
LT Domestic Cat. I	501
LT Cottage Ind. Cat. IV	2
LT Irrigation/Agrl. Cat. V	703
LT Public Lighting (Local Bodies) Cat. VI	27
HT Irrigation Cat. IV	*
HT Rural Elec. Co-op. Societies	90
Additional Subsidy to LT Agrl. Cat. V on account of annualization	22
Total	1345

Source: Power Sector Status and Tariff in Andhra Pradesh (2000-1), APERC.
Note: *Included in LT Category V.

AP TRANSCO committed to internal efficiency gains of Rs. 500 crore claiming that T&D figures were at best guestimates and as only 41 per cent of total energy purchased was being billed they could strive to increase this figure. Accordingly, the APERC set the tariff after ensuring that agricultural consumers would not be burdened. Consequently, domestic consumers had a rate shock (see Table 9.16).

Even as the APERC's decisions on tariffs appeared in the newspapers, there was a public outcry about the extent to which the tariffs were hiked particularly for domestic consumers who consume less than 200 units or kWh per month. This was further fuelled by the results of the cost to serve analysis which appeared in the tariff order and gave rise to fears that over the years tariffs for domestic consumers would be the highest closely followed by those for agricultural consumers thus reversing the traditional method which

TABLE 9.16: DOMESTIC TARIFFS

(*Ps. per unit*)

Non-Telescopic Slabs	Existing rates	AP TRANSCO Proposed rates	Telescopic Slabs	APERC Proposed rates	After Subsidy of Rs. 13450 Mn.
0-50	80	120	0-50	220	135
0-100	120	180	51-100	475	365
0-200	165	250	101-200		
0-300	210	300	201-300	635	575
0-400	290	350	301-400		
>400	340	390	>400	775	665

Source: Tariff Order dated 27.5.2000 and subsequent.

TABLE 9.17: CATEGORY-WISE REALIZATION ON COSTS

Category	SNC Lavalin FAC in Ps/Unit	APERC FAC in Ps/Unit	Average Realization in Ps/unit	% to FAC
LT Domestic	455	496	168	34%
LT Non-domestic	375	368	413	112%
LT Industrial	336	324	369	114%
LT Agriculture	329	236	18	8%
HT Industry	203	208	439	211%
HT Railway Traction	202	193	428	222%

Note: FAC: Fully Allocated Cost.

had incorporated cross-subsidies (see Table 9.17).[11] Part of the reason for the somewhat disproportionate burden on the low-end consumers was the conversion from a six-slab system to a four-slab system. Consequently, the lowest two categories changed from 0-50 and 0-100 units per month to 0-50 and 51-200 units per month. This was somewhat mitigated by a shift from non-telescopic slabs to telescopic slabs.

In addition, the annualization of the tariff adjustments by the Commission to facilitate AP TRANSCO to recover its revenues for the full year of 12 months in 10 months further increased the tariff. The GoAP quickly proposed an additional subsidy of Rs. 281.25 crore to be adjusted against the revenues from the domestic consumers and the APERC revised the domestic tariffs accordingly. Even then, the agitation took several months to die down after several ugly incidents. The chief minister of Andhra Pradesh promised the agitators that there would be a reversion to the six-slab system in the following tariff order. Subsequently, in the following year, the tariff order dated 24 March 2001 specified a reversal to the six-slab system and apart from a nominal increase for high end consumers, tariffs for domestic and agricultural consumers were left untouched. This was not surprising as local body elections in the state were scheduled for a few months later.

One of the positive outcomes on these events was the public awareness on electricity issues that it created. For the first time, the word 'T&D Losses' became a common man's terminology and the AP TRANSCO was being questioned as to why its commercial losses should be paid for by the 'paying' consumers and the government.

However, what was not highlighted in the entire exercise was an inordinately high cost of power which continues to increase (see Table 9.18). If almost 84 per cent of the total expenditure is on

TABLE 9.18: POWER PURCHASE COST TO TOTAL EXPENDITURE

Description	1994-5	1995-6	1996-7	1997-8	1998-9	1999-00 (Prov.)	2000-1 (Projected)
Expenditure	3133	3572	4280	5498	7036	7375	8366
Power Purchase Costs	1686	2105	2415	3464	5006	5746	7041
% of Power Purchase Costs to Expenditure	54	59	56	63	71	78	84

Source: Power Sector Status and Tariff in Andhra Pradesh (2000-1), APERC.

account of purchased power for which there is at present no regulatory oversight, would be possible at all to arrive at 'fair' tariffs?

9.2.5 Dealing with the Past

One of the most significant obstacles to reforms in the power sector is the handling of outstanding dues. Even while the state governments have to support the power sector in terms of subsidies in the transition period and beyond, the outstandings have become a stumbling block. The outstanding dues of the AP Power utilities to central government entities is provided in Tables 9.19 and 9.20.

In 1999-2000, while the government received a total of Rs. 278.1 crore from the power utilities the budgetary dispersal was Rs. 687.3 crore. From the year 2000 onwards the government support as subsidies to the power sector has in fact been increasing. In the Tariff Order 2000 it was Rs. 1,626 crore. However in the Tariff Order 2001 when the GoAP opted to subsidize most of the revenue gap, the subsidy came to be Rs. 1,561 crore.

Given its precarious fiscal position, the GoAP, it appears, would

TABLE 9.19: OUTSTANDING DUES PAYABLE BY A.P. POWER UTILITIES

(*Rs. in crore as on 31.1.2001*)

Corporation	Principal	Interest	Total
To National Thermal Power Corporation Ltd.	342.6	94.4	437.0
To Power Grid Corporation of India Ltd.	27.8	-8.5	19.3
To Power Finance Corporation	142.9	0.0	142.9
To Rural Electrification Corporation	–	0.9	0.9
Total	513.4	86.8	600.2

Source: Collected from Ministry of Petroleum, GoI, PFC and NTPC.

TABLE 9.20: OUTSTANDING DUES PAYABLE TO OTHER CENTRAL PSUs

(Rs. in crore as on 31.1.2001)

To Railways	25.3
Bharat Heavy Electricals Limited	288.6
Neyveli Lignite Corporation	479.9
Nuclear Power Corporation	165.3
Coal Sector*	21.1
Total	980.2

Source: Collected from Ministry of Power, Power Finance Corporation and AP TRANSCO.

Note: *All subsidiaries of Coal India Ltd and Singareni Colleries Co. Ltd.

have to continue ploughing in large subsidies to the reforming power sector even while the burden of tackling past liabilities remains.

9.3 CONCLUSION

Power sector reforms in Andhra Pradesh have had many twists and turns. Although reforms were initially proposed to bring in fresh investment, the reforms were designed to correct the many ills besetting the sector and became part of a larger economic restructuring project with the World Bank. The long transition period in the reforms and restructuring exercise has in fact been a deterrent to the reforms process itself as the financial situation of power utilities—predominantly AP TRANSCO and the distribution companies have deteriorated with the progress of reforms. And, more importantly, consumer awareness of reforms and perception of changes in a positive direction has been insignificant.

The government has a much more positive role to play in the reforms process than involving itself in the day to day operations of the power sector. To begin with, the government should take the first step towards setting a good example of paying bills; for a start a budgetary provision may be made for the payment of electricity bills in government-owned schools, health and other public institutions. Secondly, the AP TRANSCO and the four subsidiary distribution companies should incorporate professional management at the earliest. The delay in doing this has also harmed the reforms process. And thirdly, a solution needs to be found to tackle the accumulated liabilities of the Andhra Pradesh State Electricity Board which would not 'front load' the burden on the consumers of electricity in the state.

The task of public information and education should have been taken much more seriously to enable consumers to better understand the status of the power sector, the problems it suffers from and how reforms are meant to tackle these problems. This lacuna has created widespread misconceptions in the public mind on the efficacy of the reforms process.

Although there are variations in the reforms process in the power sector from state to state, the issues that need to be tackled are somewhat similar. So are the transitional issues. Therefore unless the transition is handled effectively, the reforms process itself may be jeopardized.

NOTES

1. This is very clear from the GoI Policy Document of 22 October 1991 which states:
'With the objective of bringing in additionality of resources, for the capacity addition programme in the electricity sector, Government have formulated a policy to encourage greater participation by privately owned enterprises in the electricity generation, supply and distribution field. The policy, in this regard has widened the scope of private investment in the electricity sector, and has introduced modifications in the financial, administrative and legal environment, for the private enterprises in the electricity sector towards making investments in the sector by private units attractive. Based on this policy, a scheme has been framed to encourage private enterprises' participation in power generation, supply and distribution. . . .'
2. The ESA 1948 led to the establishment of the Central Electricity Authority (CEA), a technical regulatory body attached to the Ministry of Power in the Government of India and the State Electricity Boards owned by state governments and perceived as self-regulating bodies. The central and state governments were also vested with regulatory powers.
3. The Subvention from the Government of Andhra Pradesh to APSEB was:

Year	*Amount of Subvention (Rs. in crore)*	*How made available*
1990-1	386.3	By adjusting the dues of interest on State Government loans
1991-2	695.0	By adjusting the dues on interest on State Government loans
1992-3	302.5	By converting loans to equity
1993-4	2752.5	By converting loans to equity
1994-5	9441.1	By Writing off equity

4. From Detailed Power Sector Policy Statement of the Government of Andhra Pradesh.

5. The World Bank's ten year reform programme included the following phases:

Phase I: Establishment of a new legal, regulatory and institutional framework, functional unbundling of the system, corporatization of sector entities, and removal of some of the most critical bottlenecks of the power system (1999-2001)

Phase II: Regulatory Commission and the corporate entities fully functional, privatization of part of the distribution business, partial restoration of creditworthiness of the sector, and improvements in system efficiency (2001-3).

Phase III: Completion of privatization of distribution business, consolidation of the functioning and financial performance of the new power utilities (2003-5).

Phase IV: Achievement of higher customer satisfaction through reduction of power deficit, attainment of higher quality and efficiency in electricity services, and deepening of power sector reforms to increase competition and private participation (2005-7).

Source: Project Appraisal Document on a proposed loan in the amount of US $210 million equivalent to India for a Andhra Pradesh Power Sector Restructuring Project in support of the first phase of the Andhra Pradesh Power Sector Restructuring Programme, 25 January 1999, The World Bank, Report No. 18849 IN.

6. The functions of the Andhra Pradesh Electricity Commission:
   a. to aid and advise, in matters concerning electricity generation, transmission, distribution and supply in the State;
   b. to regulate the working of the licensees and to promote their working in an efficient, economical and equitable manner including laying down standards of performance for the licensees in regard to services to consumers;
   c. to issue licenses in accordance with the provisions of this Act and determine the conditions to be included in the licenses;
   d. to promote efficiency, economy and safety in the use of the electricity in the state including and in particular in regard to quality, continuity and reliability of service and enable to meet all such reasonable demands for electricity;
   e. to regulate the purchase, distribution, supply and utilization of electricity, the quality of service, the tariff and charges payable keeping in view of both the interest of the consumer as well as the consideration that the supply and distribution cannot be maintained unless the charges for the electricity supplied are adequately levied and duly collected;
   f. to promote competitiveness and progressively involve the participation of private sector, while ensuring fair deal to the customers;
   g. to collect data and forecast on the demand and use of electricity and to require the licensees to collect such data and forecast;

h. to require licensees to formulate perspective plans and schemes in co-ordination with others for the promotion of generation, transmission, distribution and supply of electricity;
i. to regulate the assets, properties and interest in properties concerning or related to the electricity industry in the state;
j. to lay down a uniform system of accounts among the licensees;
k. to regulate working of licensees and promote their working in an efficient, economical and equitable manner; and
l. to undertake all incidental or ancillary things.

Source: *The Andhra Pradesh Electricity Reform Act 1998.*

7. See G.O. Ms. No. 11, Energy (Power III) Department, 31 January 2000—Amendments to Andhra Pradesh Electricity Reform (Transfer Scheme) Rules, 1999.
8. Such a revaluation which was undertaken as part of the first transfer scheme itself, has been challenged in Orissa as the original value alone should be applicable to tariff fixation. Though the Orissa Electricity Regulatory Commission rejected the consumers' plea, the matter is being pursued in the High Court of that state by the consumers.
9. The PPA between APSEB and M/s BPL Power Projects (A.P.) Ltd. was initially signed in 1994. Consequently, the PPA was amended and the re-stated PPA signed in January 1999 and amended again in January 2001. Consequntly, under the provisions of the APER Act 1998, the matter had to come before the APERC which could decide only after a public hearing. APERC consulted the CEA for tariff calculations under different scenarios. The issues covered in the public hearing included:
   - —Delay in construction
   - —No commitment on financial closure
   - —High cost of BPL Project
   - —Is it the least cost option?
   - — RoE and full fixed charges payable at 68.49 per cent-PLF—Why should it not be higher at 80 per cent?
   - —Why incentives at all?
   - —Why deemed generation clause?
   - —Why front-load depreciation?
   - —When D&T losses decrease we may not need the power
   - —Why forex in equity and loan?
   - —Why ESCROW as AP Genco is cheaper
   - —Force majeure conditions biased in favour of BPL
   - —15-yr contract with Gridco of Orissa would be cheaper

   The APERC accepted capital costs and tariff calculations of CEA and suggested the following:

   - —Full cost recovery at 80 per cent PLF
   - —IRR is not less than 11.5 per cent even at 80 per cent PLF

— No incentive payable on deemed generation above threshold PLF
— Reduce foreign exchange in debt
— Convert forex equity into Indian Rupees
— Interest rates need to be lowered
— Discuss despatch flexibility with AP TRANSCO
— Clarification on escrow from AP TRANSCO
— Annual return should not exceed 16 per cent

The APERC also said that the AP TRANSCO and BPL should negotiate on the above issues and the approach APERC for consent of PPA

The AP TRANSCO submitted a final response on 15.10.2001. This was also rejected by the APERC in its Order P. No. 327 of 2001 dated 6.11.2001.

Source: *APERC, O.P. No. 327/2001 dated 17.08.2001 from www.ercap.org.*

10. See: Practice Directions on Procedure for Obtaining Consent of the APERC for captive power plants, APERC No. 1, dated 08.10.1999
11. SNC-Lavelin International Inc. had prepared a Cost of Service Study for Distribution in Andhra Pradesh and submitted the document to AP TRANSCO. [Cost of Service Study, Distribution, Document 700-006-A, January 2000.] This study was based on the scanty data available as well as several assumptions in connection with coincident demand and allocation of T&D losses amongst consumer categories. Based on these estimates the study arrived at fully allocated costs that are illustrated in Table 9.17. The APERC published this table in its Tariff Order dated 27.5.2000 even though it had not used these results for arriving at their tariff decisions.

REFERENCES

Ahluwalia, Sanjeev S., 'Power Tariff Reform in India', *Economic and Political Weekly*, vol. 35, no. 8, 2000, pp. 3407-19.

Byrne, J. and C. Govindarajulu, 'Power sector reforms: Key elements of a Regulatory Framework', *Economic and Political Weekly*, vol. 32, no. 29, pp. 1946-7.

Dixit, S., G. Sant and S. Wagle, 'WB-Orissa Model of Power Sector Reforms: Cure Worse than Disease', *Economic and Political Weekly*, vol. 33, no. 17, 1998, pp. 944-9.

Godbole, Madhav, 'Future of State Electricity Boards', *Economic and Political Weekly*, vol. 32, no. 35, 1997, pp. 2222-4.

Gurtoo, A. and R. Pandey, 'Power Sector in Uttar Pradesh: Past Problems and Initial Phase of Reforms', *Economic and Political Weekly*, vol. 36, no. 31, 2001, pp. 2943-60.

Haldea, Gajendra, 'Whither Electricity Reforms?', *Economic and Political Weekly*, vol. 36, no. 17, 2001, pp. 1389-91.

Hosein, D., 'Where are the Utilities Going Next?', *Business Strategy Review*, vol. 10, no. 1, 1999, pp. 37-42.

Lovei, L., 'The Single Buyer Model', Note Number 225, *Public Policy for the Private Sector*, The World Bank, 2000.

Morris, Sebsatian, 'Regulatory Strategy and Restructuring: Model for Gujarat Power Sector', *Economic and Political Weekly*, vol. 35, no. 23, 2000, pp. 1915–29.

Ruet, Joël, 'Winners and Losers of the Reform of SEBs: An Organisational Study', Occasional Paper, New Delhi: CSH, 2001.

Sankar, T.L. and Usha Ramachandra, 'Electricity Tariffs Regulators: The Orissa Experience', *Economic and Political Weekly*, vol. 35, nos. 20 and 21, 2000, pp. 1825-33.

Sappington, D.E.M., 'Incentives in principal agent relationships', *Journal of Economic Perspectives*, 5(2), 1990, pp. 45-66.

Tenenbaum, B.W., R. Lock and J. Barker, 'Electricity Privatisation: structural, competitive and regulatory options', *Energy Policy*, vol. 20, no. 12, 1992, pp. 1134-60.

Wood, D. and D. Kodwani, 'Privatisation policy and power sector reforms: Lessons from British experience for India', *Economic and Political Weekly*, vol. 32, no. 37, 1997, pp. 2350-8.

10

Orissa Power Sector Reforms: Getting Back on Track

SIDHARTH SINHA

The Orissa Power Sector has completed three crucial stages of the reform process—corporatization, unbundling and privatization of distribution. The regulatory body has also been functioning and has issued four tariff orders so far. However, the sector is faced with a serious financial crisis. The transmission company, GRIDCO, is virtually bankrupt and the DISTCOs are incurring significant losses. The adverse impact of this is being felt on the generating companies, OPGC and OHPC. This paper analyses the power sector reform process and the performance of the power sector entities. Not surprisingly, the main problem is the management of T&D losses and the relationship between regulated tariffs and T&D losses. The only way out of the current stalemate is for the OERC, the licensees and the government of Orissa to jointly agree to a multi-year sustainable tariff schedule, a set of efficiency improvement targets and phased out annual subsidies from the government to the licensees which ensures that the licensees break even. Although not formulated in this manner the approach in Andhra has some elements of this proposal. The Maharashtra regulator also appears to be moving towards this arrangement.

10.1 POWER SECTOR PRIVATIZATION: LESSONS OF ORISSA

The objective of the Orissa Power Sector Reforms is to improve the availability and efficiency of power supply by restructuring and substantially privatizing the power sector in Orissa. The government's ultimate objective is to provide an appropriate policy environment for growth of the power sector and withdraw from it as an operator of utilities. The sector is expected to function with privately-managed utilities operating where feasible in a competitive environment under an appropriately-regulated power market.

Like other states Orissa is also faced with the problems of power shortages, imbalance between tariffs and costs and high T&D losses. However, the proportion of agriculture consumption in Orissa, at 6 per cent, is particularly low compared to other states which report agricultural sales of 30-40 per cent of total consumption. In addition hydro-power accounts for about 40 per cent of total consumption, currently at an average price of approximately 60 paise per unit.

The key principles of the Orissa power sector reform programme are:

(a) *Restructuring of the former OSEB by corporatization and commercialization*—unbundling and structural separation of generation, transmission and distribution into separate corporations;
(b) *Privatization*—through private sector participation in the hydro generation and grid corporation and privatization of thermal generation and distribution;
(c) *Competition*—competitive bidding for new generation;
(d) *Separate Regulation*—development of a power sector regulatory commission; and
(e) *Tariffs*—related electricity tariff reforms at bulk power, transmission and retail levels.

A number of critical elements of the power sector reform process in Orissa have been completed.

- The Talcher Thermal Power Station of the OSEB has been transferred to the NTPC (3 June 1995).
- OSEB was corporatized into the Grid Corporation of Orissa (GRIDCO) and Orissa Hydro Power Corporation (OHPC) (1 April 1996).
- Orissa Electricity Regulatory Commission (OERC) was established in August 1996. It has issued four tariff orders to date. Separate bulk supply, transmission and distribution and retail supply tariffs have been issued. The level of cross-subsidy in the retail tariff has been progressively reduced. Tariff levels

have been increased at about 10 per cent in successive years.

— Orissa Power Generation Corporation (OPGC) has been privatized in 1998 with transfer of 49 per cent stake to a private operator, AES, for Rs. 603 crore.

— Four distribution companies were corporatized under GRIDCO and privatized with transfer of 51 per cent stake to private operators (1 April 1999). BSES acquired three DISTCOs in the first round and AES acquired the fourth one in the second round. GRIDCO received Rs.1.6 billion for the 51 per cent stake of the distribution companies.

The most crucial element in the power sector reform process is the improvement in efficiency of the distribution function. Improvements are necessary in the quality and the maintenance of the distribution network so as to provide high quality service to the consumers. Equally important is the need for improvements in revenue generation through metering, billing and collections. Under the SEB regime the system got locked in a vicious circle of low revenue generation leading to low quality, in turn leading to low revenue. There was rampant theft in the system through 'hooking' and bypassing or slowing of meters, often with the connivance or even active participation of the SEB employees. This resulted in the so called high T&D losses. Of course, the reported T&D losses were kept low by inflating the consumption of consumers with no meters or 'defective' meters. These consumers and SEB employees benefited from the system even though the system as a whole was constantly deteriorating. The entire system was propped up by subsidies from the government and cross-subsidy by industrial consumers. However, this arrangement is no longer sustainable because the government is running out of funds to pay subsidies and industrial consumers are finding the cross-subsidizing tariffs unsustainable in a competitive environment. Many have resorted to captive power.

The reform process has two main components to improve the efficiency of the distribution system. The first is the privatization of the distribution companies. This is expected to lead to better incentives for efficiency and an overall commercial approach to management. The second component is tariff setting by an independent regulatory agency through a transparent process. Tariffs set by a regulated agency are expected to ensure that the distribution companies will cover their costs and earn a reasonable return at efficient levels of operation.

The current problem with the Orissa power sector reforms is that the expectations of efficiency improvements have not been achieved, or at least the pace of improvement is below expectation. T&D losses continue to be high. However, tariffs based on low T&D loss benchmarks are inadequate for the distribution companies to cover their costs and earn a return. Regulators blame the DISTCOs for the lack of improvement in efficiency. The DISTCOs blame the regulators for setting unrealistic efficiency targets and tariffs which are grossly inadequate to cover costs. The government, in turn, blames them both.

The net result of the stand-off is that GRIDCO is financially bankrupt and the DISTCOs are facing serious liquidity problems. The resulting losses imply that the DISTCOs are not able to pay GRIDCO for bulk supply, which in turn is not able to pay the generators resulting in a chain of payables and receivables. Losses have to be ultimately financed, but the regulator does not allow the interest on such loans to be passed through to tariffs. The regulator takes the stand that the losses would not have occurred in the first place if only GRIDCO and DISTCOs were able to achieve the efficiency targets. The DISTCOs in turn claim that given the SEB legacy they have inherited, including the manpower, they are unable to accelerate the pace of improvement in efficiency.

The power sector problems reached such a pass that AES, the private sector partner for CESCO, threatened to withdraw from the company. The OERC, fearing an emergency situation, has stripped AES of the management control of CESCO and vested it with two government nominated IAS officials.

This chapter analyses the major reasons for the stalemate in the power sector reforms process in Orissa. The only way to put the reforms back on track is for the OERC, GRIDCO, DISTCOs and the government of Orissa to arrive at an integrated plan for the sector which provides for realistic multi-year tariff increases, T&D loss reductions and subsidies during the transition period. The plan should also identify the duties and responsibility of each party in ensuring that the plan is successful. This paper provides the basic elements of such a plan.

The chapter is organized as follows. Section 10.2 analyses the OERC's approach to setting tariffs, especially the treatment of T&D losses and reduction of cross subsidies. Section 10.3 outlines the privatization processs of DISTCOs, a key feature of the reform process. Section 10.4 assesses the current financial position of the privatized DISTCOs and looks at the steps taken by them to improve their financial and operating performance. Section 10.5 assesses GRIDCO's financial performance and the attempts at financial restructuring. Section 10.6 looks at the manner in which the government has gained from the reform process but has failed to discharge its obligations. Section 10.6 concludes with some suggestions for setting the reform process back on track.

10.2 TARIFF SETTING AND T&D LOSSES

So far the OERC has issued four tariff orders incorporating the following (Table 10.1) retail tariff increases. The first two increases were by the erstwhile OSEB; the remaining four are by the OERC.

As seen from Table 10.2 there has been an attempt to rebalance tariffs and reduce cross-subsidies. The increase in domestic and

TABLE 10.1: RETAIL TARIFF INCREASES

1994-5	1995-6 (5.11.95)	1996-7 (21.5.96)	1997-8 (1.4.97)	1998-9 (1.12.98)	1999-2000 (1.2.2000)	2000-1 (1.2.2001)
Base Year	17.5%	17.0%	10.33%	9.3%	4%	10.2%

Source: GRIDCO Overview, p. 114.

TABLE 10.2: ELECTRICITY TARIFF IN ORISSA FROM 21.5.1996

	Units/ Month	1996-7	2000-1	% increase over 1996-7
Effective from		21-5-96	1-2-2001	
Domestic	100	94	150	59.57
Domestic	200	122	185	51.64
Commercial	100	210	340	61.90
Commercial	200	250	370	48.00
Small Scale Inds	20% LF	165	346.73	110.14
Med. Industries	30% LF	230	343.00	49.13
Large Industries	70% LF (HT)	248.49	310.63	25.01
Large Industries	70% LF(EHT)	273.49	297.77	8.88
Pow. Int. Industries	80% LF(HT)	268.05	293.05	9.33
Pow. Int. Industries	80% LF(EHT)	268.05	279.30	4.20
Ministeel	40% LF	238.10	366.10	53.76
Rly. Traction	40% LF	306.10	366.10	19.60

Source: GRIDCO Overview, p. 115, and OERC.

TABLE 10.3: SUBSIDY ON AVERAGE COST BASIS

(Ps. per unit)

Category of Consumers	2000-1
Domestic	182.64
Commercail	-25.85
Small Industry	16.65
Medium Industry	-14.94
Large Industries	-172.29
Railway Traction	-245.81
Mini Steel Plant	-188.35

Source: OERC, 'Role of OERC in Power Sector Reform and Tariff Setting'.

commercial tariff has been approximately 50-60 per cent as against much lower increases for the industrial sector. However, the domestic sector continues to be subsidized quite significantly by the industrial sector. For example, as shown in Table 10.3 during 2000-1 the per unit subsidy received by domestic consumers was 182 paise as against a per unit subsidy payment of 172 paise by large industry. In terms of total subsidy the domestic sector received a subsidy of about Rs. 445 crore during 2000-1 mainly from the industrial sector.

In fixing tariffs the major problem that the OERC has had to contend with is the determination of T&D losses. There are some definitional ambiguities about the T&D loss measure. The three relevant figures are input (in units), billing or sales (in units and Rs.) and collection (in Rs.). T&D loss can be defined as the difference between inputs and billing in units and expressed as a percentage of the input.

$$1 - \left[\frac{\textit{billing in units}}{\textit{input in units}} \right]$$

In setting tariffs this is the approach used by the OERC to determine the input into the system given the sales forecast. Tariffs are then fixed so as to recover the cost of power input as well as other costs. Given that only about 50 per cent of LT consumers have functioning meters a significant proportion of the billed units are estimates and not actual supply. Thus the T&D loss measure is only an estimate.

This measure of loss effectively assumes that collection is 100 per cent of billings. In the case of Orissa this is not true, especially for LT consumers. In order to account for this a different measure can be used.

$$1-\left[\frac{\textit{billing in units}}{\textit{input in units}}\times\frac{\textit{collection in Rupees}}{\textit{billing in Rupees}}\right]$$

The expression in brackets can be referred to as 'the effective percentage of input units to the system realized as cash at the average tariff realized'. This measure of T&D loss may be more relevant for measuring the efficiency of DISTCOs given the less than 100 per cent collection of billings.

In its first tariff filing in 1996 GRIDCO projected T&D losses of 42 per cent for 1997-8 as against the current level of 47 per cent. However, the OERC did not accept GRIDCO's estimate and instead appears to have based its decision regarding allowable T&D losses on the targets established by the World Bank during the initial processing of the loan. In early 1996, OSEB's system losses were estimated at 43 per cent for FY 96 and from that base GRIDCO's system loss reduction targets were fixed. Based on this, OERC decided to allow T&D losses of 35 per cent for 1997-8, consisting of 20 per cent technical losses and 15 per cent non-technical losses. During the project implementation, with the availability of detailed information and data, the 1996 base figures for OSEB's last year of operation was raised to 52-3 per cent. The figure for GRIDCO's first year of operation was raised to 50.5 per cent. Accordingly, the loss reduction targets were revised as shown in Table 10.4.[1]

However, as shown in Table 10.5, in subsequent tariff orders the OERC persisted with the 35 per cent T&D loss benchmark.

In the tariff order for 1998-9 the OERC offered to 'to treat the amount of difference between the revenue requirement calculated on the basis of GRIDCO's estimated T&D loss of 41 per cent and the revenue requirement calculated on the basis of the reasonable level of 35 per cent as has been determined by us as special category of capital'. This amount with RBI rate of interest would be considered for inclusion in the revenue requirement for tariff purposes only when GRIDCO or its successor licensees produced evidence of having reduced T&D losses to the level of 35 per cent. Additionally, if GRIDCO or its successor licensee was able to reduce the T&D loss below 35 per cent, the OERC would consider increasing the return from the existing level by one percentage point for every percentage decrease in the T&D losses.

The licensees in their tariff applications have requested OERC to permit special appropriation to cover previous losses, arising on account of their inability to achieve the T&D loss targets set by the

TABLE 10.4: T&D LOSS PROJECTION AND REVISION BY THE WORLD BANK

Loss	FY 97	FY 98	FY 99	FY 2000	FY 2003
Initial Projection	39.5%	34.8%	29.2%	24.3%	20.6%
Revised	50.5%	47%	41%	35%	22.7%

Source: GRIDCO Overview, p. 51.

TABLE 10.5: T&D LOSSES: PROPOSED AND APPROVED

Year	T&D Losses	
	Proposed	Approved
1997-8	42%	35%
1998-9	41%	35%
1999-2000		35%
Transmission	5.3%	4%
Distribution	40% (BSES)	31%
	47% (CESCO)	
2000-1		
Transmission	4%	3.7%
Distribution	38% (BSES)	31.46%
	42.66% (CESCO)	

Source: OERC Tariff Orders.

OERC. As per the Sixth Schedule past losses can be financed by way of special appropriation in any year to the extent permitted by the Commission. However, the OERC concluded, 'We are of the opinion that only where surplus of income over expenditure is available the licensee can request and commission may decide the extent to which surplus can be appropriated towards past losses in calculation of clear profit.' According to the OERC, since there was no surplus, there was no question of appropriating any amount for past losses. The OERC has also not allowed interest on bonds issued to fund losses. According to the OERC the losses were the result of inefficiencies, the cost of which could not be passed on to consumers.

Finally, in the tariff order for 2000-1 the OERC decided to leave a revenue gap which could be recovered later: The Commission felt that, 'Strict adherence to the Sixth Schedule will result in what may be called tariff shock in view of sudden rise in tariff. The industrial load growth in Orissa has been in the negative and hence full increase to fill the revenue gap may allow further shrinkage in the electricity demand.' In this tariff order the OERC also directed the licensees to carry out pilot studies for determination of technical and non-technical losses and submit the same to the Commission after which the Commission would take a view on loss reduction programme. The study is yet to be completed.

10.3 DISTRIBUTION PRIVATIZATION

The key element of the reform process is the privatization of distribution. The purpose of distribution privatization is to create the right incentives and motivations for improvements in efficiency. Privatization also insulates the organization from political interference and interventions and allows it to function in a commercial manner. Overall, privatization is expected to achieve improvement in work culture, something considered difficult under a government set-up.

The initial plan was to privatize the distribution system gradually, area by area, by the end of year 2000. The entire distribution system in Orissa is divided into 10 circles, which are further subdivided into forty-three divisions. The 10 circles were grouped into four zones, each of which was incorporated as a distribution company. The distribution company areas were selected so that they were reasonably balanced in terms of consumer mix and future growth potential.

10.3.1 Distribution Operation Agreement

The main obstacle to quick privatization was the lack of detailed and accurate information about the assets and liabilities of each circle and the fact that the new Regulatory Commission was not yet in place. Given the uncertainties about the regulatory commission and the lack of detailed information, any long term transaction with the private sector would require the Orissa government to give such guarantees and assurances to the private sector that there would be very little transfer of risk.

It was, therefore, decided to initiate the distribution privatization process through a short term distribution operations agreement (DOA) that terminated at a fixed date. By this time it was expected that the regulatory commission would be in place and would also have had an initial track record. The bulk of investments in system rehabilitation and reinforcement would be made by GRIDCO and the private company would be invited to help manage and execute GRIDCO's capital investment programme. It would be expected to provide its own staff for key management posts, the rest being seconded by GRIDCO to work for and be managed and paid by the company. The subsequent long term lease or sale contracts would shift the investment obligation from GRIDCO to the new companies. Staff would similarly be transferred from GRIDCO and become regular employees of the companies.

Four companies, that had earlier expressed interest in taking over parts of OSEB's distribution system and had signed MOUs for specific, partly overlapping areas, were invited in February 1995 to participate in structured negotiations to take over circles in the area of the proposed central company (comprising three circles: Bhubaneswar, Cuttack and Dhenkanal). After several rounds of technical offers, price bids and negotiations, one of the four companies, Bombay Suburban Electric Supply Company (BSES), was selected and a Distribution Operations Agreement (DOA) was negotiated for a period of three years starting October 1996.

However, the DOA had to be terminated prematurely only 6 months later, in April 1997, due to non-performance as per the provisions of the contract. According to GRIDCO, the main reason for terminating the contract was that BSES had consistently failed to perform. Examples of non-performance included failure to produce revenue information; to assemble basic records for billing purposes; and to produce maintenance plans.

According to BSES, the DOA had certain major drawbacks. Employees working in the central zone were not on deputation to the BSES and hence were not accountable to it. They belonged to GRIDCO and hence did not feel responsible to BSES. Before entering into the contract BSES tried to get officers and staff on deputation. But the Government of Orissa's tribunal ruled that personnel would not be placed on deputation and that they would remain employees of the government alone. In terms of investment, about 50 per cent of the meters in the central zone needed to be replaced. However, BSES was not willing to invest in improving the technical infrastructure since it only had management control but no ownership rights.

The main lesson from this experience was that privatization, to be meaningful in Orissa circumstances, had to ensure that the private sector had operational control and the responsibility for financing and implementing system investments. This would be possible only if the private sector had a majority equity stake (at least 51 per cent) in the company so as to ensure proper management and operational control, as well as the authority to make critical decisions. These features, taken together, would mean that the distribution businesses would be at least joint ventures (if less than 100 per cent of the shares were divested), in which the requisite authority and control were in the hands of the private sector, as a part of a well structured shareholders agreement. It would also mean that the regulator would make decisions concerning the licensee's performance standards and actual performance, customer relations, and tariffs. The government's role would be substantially reduced in ownership, operation, and regulation of the sector.

10.3.2 Privatization[2]

The privatization process involved sale of 51 per cent of the equity in each of the distribution companies. Up to 10 per cent of the shares in each distribution company was made available out of GRIDCO's retained stake for the benefit of employees of the particular distribution company through an appropriate scheme.

GRIDCO issued a Request for Qualification (RFQ) in November 1997. Statements of Qualification (SOQ) were received from 12 Indian and foreign companies/consortia. The Request for Proposal (RFP) documents were provided to the pre-qualified companies/ consortia in August 1998. As there were four DISTCOs for sale, the RFP indicated a preference that no purchaser take more than two, so that some degree of 'yardstick' competition could be maintained, and the perceived risks of putting 'all eggs in one basket' could be reduced.

The privatization process had to be co-ordinated with obtaining a tariff order from the OERC since this would be an important input in the financial bids. GRIDCO made a tariff application to the OERC on August 1998. GRIDCO had previously incorporated four wholly-owned subsidiaries to take over the business of distribution and retail supply; these subsidiaries had applied to the OERC for licenses. Hence, the tariff application included a proposed Bulk Supply Tariff for transactions between the Transmission and Bulk Supply Licensee and the Distribution and Retail Supply Licensees. The tariff application, a public document, was available to the bidders for their review. The tariff order was issued by the OERC in November 1998.

The Government of Orissa issued a notification in November 1998 called the Orissa Electricity Reform (Transfer of Assets,

Liabilities, Proceedings and Personnel of GRIDCO to Distribution Companies) Rules 1998. The notification contained the distribution of assets and liabilities among the four DISTCOs and GRIDCO. In this distribution GRIDCO retained 66 per cent of the share capital, 76 per cent of the loans, 60 per cent of the fixed assets and all the accumulated losses. The notification stated that the terms and conditions of the service applicable to the GRIDCO employees being transferred to the DISTCOs 'shall not in any way be less favourable than or inferior to those applicable to them immediately before the Appointed Date'.

Initially only three bids were received—from BSES Limited (BSES), the consortium of Tata Electric Companies and Viridian Group (Tata/Viridian) and the consortium of Singapore Power and Grasim Industries Limited (Singapore Power). It was clear from the Bid values that the Bids submitted by BSES were the highest for each of Wesco, Nesco and Southco.

Given that there was no bid for Cesco the Gridco board decided that Cesco be re-offered to the 7 companies and consortia that had not formally withdrawn from the sale process. Financial Bids were received from BSES and Tata/Viridian. Both bids demonstrated extreme concern as to the cash flow shortfall of Cesco in the early years of its operations. After considerable debate and discussion it was decided that the Tata/Viridian bid was subject to conditions which were either not acceptable or not achievable and that the BSES bid was so imprecise that it was not capable of proper evaluation and constituted little more than an expression of interest.

In spite of repeated attempts by GRIDCO to conclude negotiations with Tata, no agreement could be reached. GRIDCO then took up negotiations with other potential privatization partners. GRIDCO and AES/Jyoti Structures (of Mumbai), which had expressed interest but had not bid, entered into discussions and the AES consortium essentially matched Tata/Viridian's bid, without comparable conditions. Since GRIDCO had exhausted all possibilities from the competitive solicitation, this approach to the transaction for CESCO was proper and unobjectionable. AES—which also owned 49 per cent of OPGC, and was negotiating a PPA for a new IPP—would, under the Act, need OERC approval before it could also enter the distribution business. On 27 August 1999, GRIDCO announced that it had reached an acceptable agreement with the AES consortium. OERC approval was also obtained on the same date. Table 10.6 summarizes the outcome of the privatization process.

TABLE 10.6: TOTAL PROCEEDS OF DISINVESTMENT OF 51 PER CENT OF EQUITY SHARES OF THE DISTRIBUTION COMPANIES

Name of the Company	Name of the preferred purchaser	51% Equity offered for sale Book value (Rs. crore)	Winning Bid (Rs. crore)	Winning Bid per share in Rs. (Face value Rs.10 per share)
WESCO+ NESCO	BSES	58.43	88.20	15.10
SOUTHCO	BSES	19.21	28.81	15.00
CESCO	AES-JYOTI	37.09	42.00	11.32
TOTAL		114.72	159.01	13.86

Source: GRIDCO Overview, p. 62.

The privatization process was marked by insufficient investor interest, with only three bidders finally submitting bids. Frontier Economics, as a part of its evaluation of the State Power Sector Reform, carried out an Investor Survey to assess the reasons for lack of participation by several investors who had evinced interest in the early stages of the privatization process. The main reasons cited were:

— High level of losses and collection risk. Poor condition of system. Difficulty of collection for private company without political support.
— Problems of inherited staff. Difficulties of imposing own management and bringing in own employees.
— Size of system could not support estimate of fixed costs (e.g. 3-4 expatriate employees in India). Investment is too small. Lack of paying capacity among consumers.
— Tariff uncertainty. Final tariff decision based on giving all participants a little resulting in tariffs not high enough. Lack of guarantees that tariff will reward investment and efficiency gains. Never made clear what investment would be allowed to enter the asset base and be recovered.
— Poor information from Government of Orissa and GRIDCO.
— Returns too low for risk.

10.4 PERFORMANCE OF PRIVATIZED DISTCOs

At the end of almost two years of privatization the DISTCOs are showing some improvements in billings and collections. However, they continue to show financial losses. This is mainly because of their inability to meet the T&D loss targets set by the OERC. Currently the losses are being financed by stretching their payables to GRIDCO, which creates further problems for GRIDCO.

In the first year of operations—1999-2000—the DISTCOs suffered significant losses. As shown in Table 10.7 for all the four DISTCOs the Gross Margin in terms of billings was a positive Rs.141 crore, but a negative Rs. 222 crore in terms of collections. In other words, collections were insufficient to cover their power purchase bills by Rs. 222 crore.

There was some improvement during 2000-1 (Table 10.8) but collections were still inadequate to cover power purchase costs.

Table 10.9 summarizes the collection performance of the privatized DISTCOs. During 1999-2000 while LT collections increased by about 19 per cent EHT collection was down by 11 per cent resulting in an overall marginal decline in collection of 1 per cent. During 2000-1 the growth in LT collection was much higher at 26 per cent. Thus there is a definite improvement in performance during 2000-1 and it needs to be seen whether it can be sustained during 2001-2.

TABLE 10.7: FINANCIAL PERFORMANCE OF DISTCOs (1999-2000)

(*Rs. in crore*)

	CESCO	NESCO	WESCO	SOUTHCO	Total
Billing	493	329	419	220	1,461
Collection	336	244	350	168	1,098
Power Purchase Cost	465	326	345	184	1,320
Gross Margin					
Billings	28	3	74	36	141
Collections	(129)	(82)	5	(16)	(222)

TABLE 10.8: FINANCIAL PERFORMANCE OF DISTCOs (2000-1)

(*Rs. in crore*)

	CESCO	NESCO	WESCO	SOUTHCO	Total
Billing	584	365	472	235	1,656
Collection	437	276	364	190	1,267
Power Purchase Cost	526	352	367	194	1,439
Gross Margin					
Billings	58	13	105	41	217
Collections	(89)	(76)	(3)	(4)	(172)
Other expenditure as approved by the regulator	180	113	127	102	522

TABLE 10.9: DETAILS OF COLLECTION OF DISTCOs

(Rs. in crore)

All DISTCOs	1998-9	1999-2000	Growth (%)	2000-1	Growth (%)
LT	253	302	19%	380	26%
HT	327	326	0%	383	17%
EHT	528	470	-11%	503	7%
TOTAL	1,108	1,097	-1%	1,266	15%
Input (MU)	10,056	9,989	-1%	10,857	9%

TABLE 10.10: DISTRIBUTION LOSSES

	2000-1	1999-2000	1998-9	1997-8
Input (MU)	10,856	9,887	10,056	9,646
Billing (MU)	6,080	5,679	5,431	5,571
Billing/Input %	56%	57%	54%	58%
Billing (Rs. in crore)	1,656	1,493	1,336	1,375
Collection (Rs. in crore)	1,267	1,097	1,108	1,112
Collection/Billing %	77%	73%	83%	81%

The main reason for the continuing losses is that distribution losses continue to be significantly higher than the benchmark set by the OERC for tariffs. As shown in Table 10.10 the billing to input ratio is about 56 per cent implying that almost 44 per cent of the input power is not billed. In addition collection to billing is only about 77 per cent as against the 100 per cent collection assumed by the OERC.

The problem of T&D loses is likely to be difficult to resolve because, as shown in Table 10.11, domestic consumers are predominantly rural. Almost two-thirds of domestic consumption is rural and the collection to billing ratio for rural consumers is only 40 per cent as compared to 63 per cent for urban consumers.

10.4.1 Progress on Metering

Table 10.12 shows the progress in metering since privatization.

Therefore, while there has been improvement in metering, almost half the consumers are still effectively unmetered. It is for this reason that billing numbers tend to be unreliable and greater reliance has to be placed on collection figures. There is a major problem of defective meters with almost 35 per cent of installed meters being defective. According to the distribution companies the pace of

TABLE 10.11: BILLING AND COLLECTION OF DOMESTIC CONSUMERS

	Consumers	Billing		Collection	Collection / Billing %
		MU	Rs. in crore	Rs. in crore	
CESCO					
Urban	1,67,528	340	65	45	69%
Rural	3,82,931	616	104	38	37%
NESCO					
Urban	52,357	76	14	13	93%
Rural	189,390	260	44	20	45%
WESCO					
Urban	109,730	219	41	14	34%
Rural	146,457	247	42	11	26%
SOUTHCO					
Urban	115,754	162	28	21	75%
Rural	187,926	198	33	20	61%
TOTAL					
Urban	445,369	797	148	93	63%
Rural	906,704	1,321	223	89	40%

TABLE 10.12: METERING PROGRESS

All DISTCOs	1 Apr 1998	1 Apr 1999	31 Mar 2000	31 Mar 2001
Consumers		14,16,686	16,00,551	17,26,956
Meters	10,35,503	11,18,084	12,51,314	13,89,588
Additions		82,581	1,33,230	1,38,274
Defective meters	5,21,153	5,59,515	5,21,714	4,75,290
Good meters	5,14,350	5,58,569	7,29,600	9,14,298
% of consumers		39%	46%	53%
% of all meters	50%	50%	58%	66%

metering has been slow partly because of World Bank procedures for procurement of meters. These is being taken care of and the pace of metering is likely to increase in the coming months.

10.4.2 Initiatives for Improving Efficiency

The DISTCOs have taken several initiatives to improve collection and reduce losses. The most important of these has been changing the work culture to introduce a commercial orientation among the employees. This continues to be the most difficult challenge faced by the distribution companies. Various initiatives have been taken to incentivize improvement in performance ranging from monetary incentives to motivation through empowerment.

CESCO has taken some bold steps including open-market recruitment of professionals (about 50) into their 'business cadre', assigning accountability and powers to managers and workers and taking strong disciplinary measures for non-compliance. These measures have met with resistance from vested interests among employees and other interest groups. Not many of the existing employees have exercised the option to join the alternate cadre which provides higher remuneration but no government job security. In the case of BSES companies only four executives have been brought from outside. The attempt is to manage the process by 'reforming' the existing employees.

Other steps to improve efficiency include:

- — Disconnections to force consumers to pay their bills.
- — Bijli Adalats and consumer camps to provide speedy redressal of complaints.
- — Establishment of vigilance wing to detect theft and malpractice.
- — Administrative action against erring employees.
- — Establishment of customer care centres.

10.4.3 Village Committees[3]

By far the most significant and innovative activity is the formation of Village Committees. The Village Committees have been instrumental in rural metering, billing and collections. There had been instances of village committees being formed during RIAP implementation. After privatization WESCO initiated the process with the help of faculty from the local Xavier Institute of Management (XIM). This has now spread to the other BSES companies and CESCO has also followed. Details of this 'microprivatization' are provided in the Annexe.

Village Committees address the inherent difficulties in serving widely dispersed rural loads. The Village Committee is a committee of consumers from the village with participation of field staff of the utility with the objective of facilitating a single point contact between the utility and the consumers of the village. In the first phase this village committee facilitates interaction, grievance redressal, bill distribution, metering and cash collection. A pilot project was launched in August 1999 with 100 villages in WESCO. There was substantial improvement in collections (almost 100 per cent although on a low base) and addition of new consumers for the utility. After this success the scheme was extended to an additional 3,000 villages in the NESCO and WESCO areas. In 1,000 villages, the second phase of the initiative has been launched whereby greater accountability is proposed to be transferred to the Committee by making it aware of the input at transformer level and actual billing being achieved from the village. All the consumers in the village are proposed to be metered in this phase. In the third phase it is expected that energy accounting will be entirely transferred to the village committee whereby they will perform like a cooperative, i.e. they will be billed based on transformer readings and entire responsibility for collections will be transferred to them. At that stage the utility may approach the OERC for a separate tariff for these consumers.

NESCO and WESCO have contracted out the process of formation and initial management of the process to XIM, Bhubaneswar, with participation from the utility. CESCO has taken the approach of introducing the same process through its own staff to increase ownership and ensure sustainability. Such committees are operational in more than 100 village in the CESCO area now.

Table 10.13 shows the financial benefits of the microprivatization initiative in NESCO and WESCO. There are significant incremental benefits even after the writing off the entire expenses of the microprivatization. There are proposals to expand this programme through franchizees. These proposals are still being developed.

10.5 GRIDCO's FINANCIAL PERFORMANCE

During the three year period, from the corporatization of the OSEB into GRIDCO (1996-7), till the privatization of the DISTCOs (April 1999), GRIDCO suffered a total loss of Rs. 765 crore before interest and Rs.1,019 crore after interest. The bulk of the losses before interest can be explained by GRIDCO's inability to achieve the T&D loss target of 35 per cent set by the OERC. GRIDCO's actual T&D losses during this period were about 45 per cent. If GRIDCO had been able to achieve the target of 35 per cent it would have resulted in additional revenue of approximately Rs. 720 crore, with very little additional costs. The additional revenue would have almost wiped out the losses before interest. With no losses, the

TABLE 10.13: FINANCIAL BENEFITS OF MICROPRIVATIZATION

WESCO (Rs. in lakh)	1999-2000	2000-1	Growth (%)
Domestic/Commercial Collection (Non-Project Area)	543.24	590.94	8.78
Domestic/Commercial Collection (Project Area)	176.76	303.06	71.45
WESCO Total	720.00	894.00	24.17
Incremental collection		110.78	
Cost of microprivatization		89.04	
Incremental benefit		21.74	
NESCO (Rs. in lakh)	**1999-2000**	**2000-1**	**Growth (%)**
Domestic/Commercial Collection (Non-Project Area)	196.54	287.78	46.42
Domestic/Commercial Collection (Project Area)	120.46	281.22	133.46
NESCO Total	317.00	569.00	79.50
Incremental collection		104.84	
Cost of Microprivatization		67.46	
Incremental benefit		37.38	

Notes: 1. All collection figures are for the period April to November.
2. Divisions covered for WESCO include BED, Bargarh, DED Deogarh, RED, Rourkela and SED, Sundergarh.
3. Divisions covered for NESCO include AED, Anandapur and JRED, Jajpur.
4. Incremental collection is the excess of actual collection for the project area over the hypothetical (or normal) collection assuming the non-project area growth rate for the project area.
5. The above calculation treats the entire costs of microprivatization as revenue expenses. There are several other long term benefits of the project, e.g. solving of long pending customer complaints, opening of lines of communications with village consumers, training inputs given to WESCO/NESCO staff and the education of consumers regarding their rights and obligations.
6. This also assumes that the entire cost is incremental. This may not be true since the DISTCOs would have to incur some costs to get even the normal collections.

interest burden would also have been much lower since the bulk of the additional loans were taken to finance losses and were not allowed to be passed through to tariffs by the OERC.

Therefore, the main cause of GRIDCO's financial problem was its inability to meet the metering, billing and collection performance targets set by the OERC. As already pointed out, the OERC based its performance benchmarks on the World Bank's assessment of T&D losses. The World Bank's June 1988 Supervision Report asks the Orissa government to consult with the OERC on the issue of pass-through of prior years' losses, 'Given that we all so severely underestimated GRIDCO's system losses in 1996 and set unachievable performance targets, the prima facie justification of such action appears to be solid.'

The main reason for the poor performance has been the lack of a system of incentives and disincentives for ensuring managerial accountability for energy sale and revenue collection. Apart from this, other important reasons for inadequate progress were the lack of a management information system and weak commercial and accounting systems in the erstwhile Orissa State Electricity Board. After unbundling and corporatization a large amount of advisory and implementation support has been provided to GRIDCO. In order to address the weaknesses in the commercial system, GRIDCO was provided support, both for developing a codified set of commercial practices as well as implementation of the same at the field level. Manuals for commercial practices like new consumer registration, billing, revenue collection, credit control, temporary connections, regularization, consumer complaint redressal, etc., were prepared and were put into implementation in most areas within GRIDCO. A detailed management information system was prepared to address the lack of availability of management information of good quality for the information and action of the top management. However, the operational systems and mechanisms have been difficult to change under its public sector status and despite the large quantum of support provided by DFID the improvements were minimal and the financial situation deteriorated rapidly.

With the separation of the DISTCOs in 1999, GRIDCO has shown profits before interest of Rs. 80 crore in 1999-2000 and Rs. 23 crore in 2000-1. However, as a result of the distribution losses prior to privatization of DISTCOs and the continuing failure of privatized DISTCOs to make timely payments for power purchases GRIDCO's financial situation continues to worsen. As on 1 April 2001 GRIDCO owed generators about Rs. 1,240 crore and total borrowings stood at Rs. 2,750 crore.

10.5.1 GRIDCO Financial Restructuring Plan (FRP)

The Government of Orissa submitted a Financial Restructuring Plan (FRP) for GRIDCO in October 1999 to the Ministry of Power, GoI, soon after the process for privatizing the DISTCOs had been completed. The plan has been under discussion since then while the condition of GRIDCO has continued to worsen. Recently the plan was submitted to the OERC for approval.

The FRP presented to the OERC by GRIDCO essentially involves conversion of loans from financial institutions and payables to generators into tax-free bonds:

— Rs. 600 crore dues to central generating stations to be securitized by issue of tax-free bonds.
— Rs. 800 crore of REC and PFC loans to be discharged by issuing tax-free bonds with state government guarantees.
— Rs. 560 crore due to state-owned generating companies (OHPC and OPGC) would be partly managed through the issue of tax-free bonds with a state government guarantee (Rs. 360 crore) and the balance would be paid out of soft loans receivable from the World Bank.
— Rs. 200 crore loan from the WB for managing deficits until the turnaround of GRIDCO.

The OERC held hearings on 8 November 2000 during which various affected parties expressed the following views:

— OPGC stated that the issue of bonds by GRIDCO to OPGC was not an attractive proposition. It also maintained that the waiver of DPS and penal charges on non-payment of dues was not an option it was considering.
— OHPC complained that despite being the cheapest power supplier, GRIDCO was honouring liabilities of other suppliers like NTPC and OPGC prior to paying OHPC dues. The total outstanding as on 31 October 2000 stood at Rs. 400.40 crore, and since there was neither any inflow of funds from GRIDCO nor any commitments of early settlement of dues, the FRP would put additional burden on the finances of OHPC.
— SOUTHCO disputed the loss reduction figures quoted by GRIDCO, and offered the following loss reduction targets: 3 per cent p.a. for the first 3 years and 2 per cent p.a. in the fourth and fifth years. (The FRP assumed a decrease in distribution loss percentage of 5 per cent p.a. for the first 3 years and 2.5 per cent for the subsequent years till the overall loss percentage reaches 20 per cent.)

— WESCO and NESCO held similar views on the FRP.
— CESCO held that the loss reduction programme envisaged by GRIDCO was too ambitious. It contended that if all the assumptions taken by GRIDCO were fulfilled, CESCO might be able to achieve a loss reduction of 2.5 per cent p.a.
— The Government of Orissa held the view that given the current financial situation of Orissa, it would be extremely difficult for it to infuse any equity into GRIDCO or to intervene financially.

According to the OERC the financial insolvency of OERC was largely the result of inefficiencies in operations, especially its inability to meet the T&D loss targets. The commission concluded by giving no commitment on future tariffs and rejected the proposal for a multi-year tariff.

The OERC order on the FRP clearly shows that the FRP has serious problems and in it current form is not sustainable. The biggest problem is with respect to T&D losses. Unless this problem is sorted out between the regulator and the DISTCOs the FRP will not provide a long-term solution.

10.5.2 Role of Government in the Reform Process

The Orissa government played a pioneering and courageous role in accepting and initiating the power sector reform process. However, it appears to have decided that with the initiation of the reform process it has no further financial responsibilities towards the sector. This is simply not possible because the sector is essentially non-viable given the high level of T&D losses.

The Orissa government has benefited financially from the reform process in the following ways:

— At the time of corporatization of GRIDCO the OSEB assets were revalued by Rs. 1,194 crore. In return the government adjusted subsidies and electricity charges payable to GRIDCO of Rs. 340 crore.
— In 1998 the state government sold its 49 per cent stake in OPGC to AES for Rs. 603 crore. Out of this amount Rs.120 crore was lent to GRIDCO and OPGC was allowed to retain Rs. 98 crore. The balance Rs. 385 crore was available to the government. A precondition for the sale was that GRIDCO

should liquidate their dues to OPGC. To comply with this requirement, GRIDCO raised Rs. 300 crore from the market at a coupon rate of 14.5 per cent and paid off OPGC's dues. (At that time GRIDCO had pleaded that during the next 18 months they would be able to pay only 75 per cent of OPGC's bills and the government had agreed to subsidize GRIDCO to the extent of 25 per cent, but till date the government has not paid any subsidy on this count to GRIDCO. The subsidy claim submitted to the government on this count is Rs. 107.44 crore.)

— There was an upvaluation of Rs. 500 crore when the Hydel assets of the state government and OSEB were transferred to the OHPC. As a result of the upvaluation the cost of hydel energy went up from 20.52 paise/unit to 38 paise per unit. One of the results of this upvaluation was that the Government was able to raise some funds for the cash starved Upper Indravati hydel project. It also enabled OHPC to consistently make a profit.

While the government has benefited financially from the reform process it has not fulfilled its obligations to the sector in several instances.

— Government subsidy to the sector has stopped since the reform process was initiated. The privatized DISTCOs have submitted subsidy claims, approved by the OERC, for rural electrification amounting to Rs. 23.23 crore. The government is yet to disburse this amount.

— State government departments and PSUs owed the DISTCOs Rs. 144.8 crore as on 31 March 2001 for power consumed after 1 April 1999. According to the statutory Transfer Scheme GRIDCO is due to receive Rs. 113.75 crore from the state government and Rs. 112.03 crore from the PSUs towards power consumed by them during the period 1 April 1996 to 31 March 1999. This totals to Rs. 225.78 crore and is still to be received by GRIDCO.

— The state government has defaulted in passing on to GRIDCO and the distribution companies their part of the World Bank loans of US $350 million. As on 30 April 2001 GRIDCO was owed a total Rs. 680 crore. Several times in the past, the World Bank had expressed unhappiness at the state government's withholding of Bank funds which were meant to be invested in the transmission and distribution system, resulting in an improvement in the quality of the electric supply. When matters did not improve, the Bank suspended the flow of World Bank funds to the state.

— The OERC has been functioning with only two members for a long period of time and the government has been unable to appoint a third member.

— The Orissa Act provides for a mechanism whereby funding of all reasonable expenditure of the OERC should not be subject to the discretion of the government. The OERC is required to submit a statement of estimated expenditure to the government for the ensuing year by December of the preceding year. This is then tabled in the state legislature. The expenditure is funded out of the consolidated fund of the state and does not require approval of the assembly by voting. This is a privilege extended only to institutions where the government feels it is crucial to avoid political influence, such as the state judiciary and the Public Services Commission. In turn, the license charges stipulated by the OERC for the transmission and distribution licensees are passed on to the government. The government budget for the OERC has decreased from Rs. 150 lakh in 1998-9 to Rs. 85 lakh for 2001-2. The Commission has not been getting the amount of funding requested. The government receives Rs. 2.5 crore as license fees from the four distribution companies and GRIDCO at the rate of Rs. 50 lakh per license.

At the minimum the state government needs to fulfil the following obligations—subsidy for rural electrification, clearing government departments and PSU receivables, passing through World Bank loan amounts and extending full support to the OERC. In addition the government is ultimately responsible for the long-term viability and success of the sector and needs to take suitable steps to put the reform process back on track.

10.6 CONCLUSION: GETTING BACK ON TRACK

At the end of over five years of the reform process in Orissa, the transmission and distribution entities are financially insolvent. The privatized distribution companies do not have adequate cash flows to even pay salaries. AES, the private partner in CESCO, has effectively withdrawn from management of the company following the OERC Order which vests management control of CESCO in a government nominee.

The main source of the problems is the gap between the efficiency and cost parameters underlying tariffs and those achievable by the licensees, and the government's inability to bridge the resulting cash deficits. Resolution of this problem is crucial to the viability of the state power sector. It is, therefore, a pre-requisite for taking up proposals for the financial restructuring of GRIDCO. Stakeholders will be willing to participate in a financial restructuring exercise only if there is a reasonable assurance of financial viability.

Not much seems to have been achieved, in terms of efficiency improvements, during the period 1996-9, i.e. the period from corporatization and unbundling to the privatization of the distribution companies. While considerable work may have been done by the consultants in terms of developing systems and procedures and building databases it does not seem to have translated into concrete improvements in efficiency. This is not surprising since changes in systems and procedures can achieve little if not accompanied by corresponding changes in work culture and motivation. Post-privatization there is a definite attempt to change the work culture and employee motivation. Various initiatives taken by the distribution companies to incentivize improvement in performance ranging from monetary incentives to motivation through empowerment are yielding mixed results. These measures have met with resistance from vested interests among employees and other interest groups.

In tariff setting the most critical decision of the OERC has been a benchmark T&D loss level of 35 per cent. This was reduced to 34 per cent in the last tariff order. The 35 per cent benchmark was based on World Bank estimates and targets. However, even after the World Bank had accepted that its earlier loss figures were grossly underestimated the Commission has refused to alter its benchmark. It has refused to accept DISTCOs' T&D loss figures because they are not backed by adequate data. However, adequate data is unlikely to be available given the current low levels of metering. In view of the financial losses of the DISTCOs it is clear that the actual T&D losses must be certainly higher than the benchmark established by OERC.

This gap between actual T&D loss and that allowed by OERC in the tariffs were apparent at the time of privatization. In the filing for the tariff order passed just prior to privatization GRIDCO had claimed an actual T&D loss figure of 46.6 per cent and a projected figure of 41 per cent. However, the OERC based its tariff order on a T&D loss of 35 per cent. At the time of privatization BSES in its Request for Proposal (RFP) had clearly mentioned that it found the loss levels projected in the Information Memorandum as optimistic and difficult to achieve. However, during the privatization process there does not appear to have been any discussions about the magnitude and financing of the inevitable cash deficits. Financing concerns were expressed only at the time of the CESCO rebid. AES asked for and received working capital financing for taking over CESCO. Overall, it appears that the privatization process was pushed through in haste without fully thinking through the implications. On the other hand, if these questions were seriously pursued it is quite likely that the deal would not have been consummated in its existing form.

The only way out of the current imbroglio is to carry out the planning exercise that was not done at the time of privatization. This will need to be a three-way exercise across the licensees, OERC and the Government of Orissa. This exercise rests on the following principles:

1. MULTI-YEAR TARIFFS TO ENABLE LICENSEES TO PLA EFFICIENCY IMPROVEMENTS AND INVESTMENTS

The starting point of the exercise is the acceptance of the principle of multi-year tariffs. The Financial Restructuring Proposal submitted by GRIDCO proposed a multi-year tariffs. The World Bank has also suggested that the OERC move to a system of multi-year tariffs. (Aide memoire January 8, 2001). 'In view of the time required for the utilities to turn around and the need for enhanced regulatory clarity for financing losses during this transition, it would be advisable that the utilities submit and the Commission considers a multi-year performance based tariff filing.' However, in its Order on the Financial Restructuring Proposal submitted by GRIDCO the OERC has rejected the idea of a multi-year tariff. According to the Order, 'We are not in a position to give clearance and commitment for future tariffs, as these will be dealt separately on a year-to-year basis in accordance with the OER act.' Multi-year tariffs are essential to enable the various players in the Orissa power sector to plan for the future with some degree of certainty. If necessary, the government could consider amending the OER Act to enable the OERC to issue multi-year tariffs.

Multi-year tariffs will also involve pass-through of power purchase costs. So far the OERC has only allowed pass-through of fuel costs but not other costs. In a multi-year tariff these other costs which are beyond the control of the licensees will also have to be passed through.

2. TARIFFS TO BE BASED ON STANDARD COSTS AND EFFICIENCY PARAMETERS

The multi-year tariffs should be based on standard costs and efficiency parameters. The standards are not hypothetical but based on the actual technologies in use. These standard costs and efficiency parameters would serve as targets which the DISTCOs would be required to achieve in a time-bound manner. These standards will, therefore, be set by the OERC after negotiations with the DISTCOs and using inputs from independent experts. Ultimately, the DISTCOs will have to agree to the achievement of these targets.

The tariffs should also provide for a phased reduction and ultimate elimination of cross subsidies. In case the government wants to provide subsidies to certain segments these should be clearly targeted and paid for separately by the government. This could be achieved through lifeline tariffs limited to consumers with a certain basic level of consumption.

3. LICENSEES WILL HAVE TO COMMIT TO A BINDING MULTI-YEAR PROGRAMME TO ACHIEVE THE STANDARD COSTS AND EFFICIENCY PARAMETERS

Licensees will have to provide a multi-year schedule of progress towards achieving the standards on which the tariffs are based. For this schedule to be meaningful the efficiency parameters should be measurable. This requires a level of metering, not necessarily 100 per cent, which will provide reasonably reliable T&D loss data. This schedule will also have to be negotiated between the OERC, the licensees and the government. The negotiations can draw upon independent experts.

4. NEED FOR SUBSIDY DURING THE TRANSITION PHASE TO BRIDGE THE GAP BETWEEN TARGETS AND ACTUALS

Given the multi-year tariff schedule and the multi-year programme of efficiency improvements the cash deficits of the licensees can be worked out. The cash deficits represent cost of the gap between standard costs and efficiency parameters and the actual costs and efficiency parameters. This cash deficit will have to be subsidized by the government. Other alternatives such as creation of regulatory assets are unlikely to be credible. The government could levy a special temporary tax to finance this subsidy.

The final outcome of this exercise will be a tariff plan from the OERC; an efficiency improvement plan from the DISTCOs; and a subsidy commitment from the government which together ensures break-even for the licensees. This will constitute a binding agreement among the OERC, DISTCOs and the government. An essential element of the plan would be a clear statement of the commitments by the licensees and the government for achieving efficiency targets and, ultimately, the viability of the sector. The job of the OERC would be to enforce these commitments.

The Andhra reform process appears to be following some elements of the process described above. The Andhra tariff orders are based not on actual costs but on some standards of efficient costs. For example, the tariff orders provide for an 'efficiency gain' of Rs. 500 crore. It also provided for an initial government subsidy of Rs. 1,345 crore which was later increased to Rs. 1,626 crore after the government requested a decrease in tariffs following public and political agitations in response to the original tariff order. In the tariff application for 2001-2 APERC estimated a loss of Rs. 1,024 crore for the year 2000-1 after taking the government subsidy into account. The APERC proposed to carry forward only an amount of Rs. 90 crore of this loss 'which is on account of factors beyond the reasonable control of the Licensee', through a special appropriation in the annual revenue requirement for FY 2001-2. The balance uncovered financial loss of Rs. 934 crore is to be compensated by the Andhra Pradesh government as subvention as per the provisions of Section 27(1) of the Reform Act.

Thus in the 'AP approach' the tariffs are based on some (undefined) efficient costs and the deficit is subsidized by the government. This needs to be accepted openly and be accompanied by a definite programme of loss reduction and the phasing out of subsidies.

The Maharashtra regulator also appears to be preparing to move to the kind of arrangement proposed above. According to press reports, MERC has commissioned three studies on cost of subsidies in the state power sector, benchmarking performance factors in generation and distribution, and share of electricity cost and power consumption in the state agricultural sector. According to the press report 'MERC is looking towards manifold usefulness of this study such as providing reasonable performance indicators, which could be imposed as targets to the utilities and facilitate the dialogue between the utilities, the regulator and the consumers.'

NOTES

1. The World Bank's June 1988 Supervision Report asks for the Orissa government to consult with the OERC on the issue of pass-through of prior years' losses, 'Given that we all so severely underestimated GRIDCO's system losses in 1996 and set unachievable performance targets, the prima facie justification of such action appears to be solid.'
2. This section is based on 'Orissa Power Sector Reform Project, Institutional Strengthening Project (ISP)', *End of Project Report*, Price Waterhouse Coopers, July 2000.
3. This section is based on material provided by Professor S. Govindarajan, Xavier Institute of Management, Bhubaneswar.

11

Developing Business Competence: A Model Based on Viable Systems

D.V. RAMANA

Post 1991, major reforms have been initiated by the central government. The essential thrust of the government's policy has been to reduce its presence in the business sector and in the process reduce the overall government spending. In the same spirit, the state governments also decided to reduce their scope of business activities. One such decision was to initiate reforms in the power sector and privatize the same.

This paper presents an alternative approach for strengthening the power sector by addressing the internal environment and suggesting a model based on the principles of cybernetics. It begins by bringing out some of important features of the power sector, explains the experiments and experiences under the Orissa model, introduces the concepts of the new model, and finally presents the model using one of the distribution companies, viz., CESCO.

11.1 TYPICAL NATURE OF THE INDUSTRY

Some important features of the business, which put the power sector on a different pedestal, are given below:

(1) *Non-reflection of actual losses*: Since most of the utilities do not have a proper metering and billing system, the actual loss of the power in the system is not reflected. One such example was the case of Orissa prior to the commencement of the reform process; OSEB claimed that the T&D loss was in the range of 20-5 per cent. Later the loss was found to be more than 45 per cent.[1] Even today, the distribution companies are not in a position to quantify the actual loss.

(2) *Performance of unit depends on the composition of customers*: Unlike most of the other businesses, the performance of a unit of the distribution company depends on the profile of the customers. In case a unit has a large component of domestic customers, the performance on all counts will be much below the average performance of the company. Table 11.1 demonstrates the same using two divisions of CESCO.

(3) *All consumers are not customers*: A large chunk of the power lost in the system is attributed to the non-technical losses. The word non-technical loss is a decent substitute for power theft. It has been established that many use power without being billed. This happens with and without the support of the employees of the distribution company. This may take the shape of non-billing, bills not distributed, wrong reading of meters, or tampering with the meter.

(4) *Licensee proposes commission disposes*: During the last three years of reforms in Orissa, it has been observed that the regulatory commission has not accepted most of the proposed expenses and losses but gone on stressing the issue of reduction of T&D losses. As the licensees were not able to achieve the targets fixed by the commission, the losses went on increasing. Table 11.2 shows the gap between the proposed and allowed expenses and losses.

TABLE 11.1: COMPARATIVE PERFORMANCE (MAR 2001)

	Billing as % of Input	Collection % of Billing	Customer Profile
Salepure	55%	60%	Dominated by domestic customer
Talcher	66%	109%	Domestic and large industry

Source: CESCO Revenue Reports.

TABLE 11.2: T&D LOSSES: PROPOSED AND APPROVED (ORISSA)

	Proposed	Approved
1999-2000		
Transmission Losses	5.30%	4.00%
Distribution	40-7.00%	31.00%
2000-1		
Transmission Losses	4.00%	3.70%
Distribution	38-42.60%	31.46%

Source: Tariff Orders issued by the OERC.

TABLE 11.3: LOSS AFTER BILLING

	Division A		Division B	
	Input	Billing	Input	Billing
LT	40	24	60	36
HT/EHT	60	54	40	36
Loss		22%		28%

So part of the losses reflected by the companies may be attributed to the non-allowance of the expenses and losses coupled with non-acceptance of the proposed tariff rise.

(5) *Profit (loss) decreases (increases) with increase in sale volume to the low-tension consumers*: It is important to note that losses (T&D) vary with the consumer mix. A division with higher LT consumers will have higher losses. Table 11.3 demonstrates the same.

The total loss in the system increases from 22 per cent to 28 per cent because the proportion of LT consumers goes up. So other things remaining constant, loss increases with the increase in sales to the LT consumers.

11.2 EXPERIMENTS AND EXPERIENCES OF ORISSA's MODEL: AN ANALYSIS

To address the problems of the power sector, the Government of Orissa decided on a comprehensive restructuring of the entire sector. Some of the steps taken were as follows:

— Unbundling;
— Devolution of Revenue Management; and
— Corporatization and Privatization.

The following paragraphs explains the principles underlying these steps and experiences thereof.

11.2.1 UNBUNDLING

State Electricity Boards (SEBs) were established as vertically integrated organizations to undertake generation, transmission, and distribution of power. The unbundling of the industry is based on the realization that it is possible and desirable to separate the transportation from the thing transported. In other words 'Electricity' as a product can be separated commercially from 'Transmission' as a service.[2] This shift in the understanding of the electricity industry may be attributed to the changes in generating costs during the 1980s.[3] These changes made the large customers think about building their own plants. All these changes, along with the shortage of public capital, made governments across the world encourage private participation in the public utilities industries in general, and the power industry in particular.

The Orissa Electricity Reforms Act (OERA) prescribed the method by which the power sector would be restructured. The Act allowed the transfer of assets, liabilities, and the staff and statutory obligations of the OSEB to the successor companies. As a consequence, the Grid Corporation of Orissa Limited (GRIDCO) and the Orissa Hydro Power Corporation (OHPC) were registered as government companies. Accordingly the following steps were taken.

Though, OHPC and GRIDCO began operation on 1 April 1996 as government-owned entities, corporatization agreements were signed with the Orissa government. Under the agreement, the government reiterated its commitment to distancing itself from their operation and management and promised to give them autonomy as commercial organizations. While the OHPC continues to be a government company, GRIDCO has taken long strides in the direction of full privatization.

Will the change in structure lead to higher efficiency and ensure decent competition? Viewpoints are divided on this issue. Experience differs from country to country and depends on the purpose for which the change has been initiated. According to Surrey,[4] for the developed countries the restructuring in the industry has been initiated to obtain cost savings in generation, whereas, for developing countries it helps in securing investment funding from the international investors. However, it is important to note that the latter's interest is not compatible with competition: international investors will be interested in high levels of profits.

Orissa has had mixed experience. Unbundling demonstrated that the business drivers of these three activities are different and require different sets of skills to manage them. It offered an opportunity to understand each activity as a separate business and manage them efficiently. Such unbundling helped in concentrating on a particular aspect of the value chain. It has shown that with good management it is possible to improve the performance of a particular activity.

For example, the transfer of the generation business to the NTPC and the OPGC has resulted in enormous improvement in the performance of the generating stations. It clearly shows the performance can be improved even without passing on the ownership to the private parties. The unbundling has also resulted in the increase of tariff for the ultimate consumer. Post-reforms, the state has profit-making generating companies side by side with loss-making transmission and distribution companies.

11.1.2 Devolution of Revenue Management

Approaches to reforms in the public sector have traditionally supported the idea of a strong centralized executive office, which would serve to rationalize the public policy processes and adopt a system-wide perspective. However, the public sector experienced unprecedented levels of change during the 1980s and the 1990s. Such pressures came from the entire range of stakeholders, viz., from governments (both central and state), and from the local consumers and communities, to develop a 'public service' orientation, involving the provision of demand-led services and be responsive to the needs of those being served.[5]

The same period has also seen the emergence of a populist search for managerial and corporate improvement largely based on the models adopted by the private sector enterprises. This resulted in a move from traditional bureaucratic modes of administration to more flexible and customer-oriented management.[6] A core element of this new managerialism is the adoption of decentralized budgetary management systems with managers at the lowest level having more authority over resources; greater scope for decisions; and accountability for achieving agreed targets.[7]

More recent reforms have abandoned hierarchy in favour of market-driven management systems, which reflect private sector management practices and values.[8] These practices and values were developed and proposed for addressing the inefficiencies of the public sector by Osbrone and Gaebler (1992).[9]

In Orissa, along with the unbundling of the business, efforts were made to decentralize the revenue management. A conscious effort was made to improve the commercial function of the company. The OSEB was managed principally by engineers, therefore, the management had a strong technical bias and it was standard practice for engineers to undertake commercial, administration and other non-engineering jobs. One of the major organizational changes introduced during the early days of the reforms was to bifurcate the field level activities into engineering and commercial functions. The commercial function comprised of revenue management, accounting, and customer management, while the engineering function comprised of system operation, repairs and maintenance, and other technical activities. Coupled with the bifurcation, commercial procedures were implemented. This small intervention improved the business environment at the division level.

The concept of profit-centre accounting was introduced in the distribution companies. Under the new practice, each division was treated as a profit-centre and was required to earn profit. Each division was put in charge of an executive engineer (re-named as divisional manager to drive in the importance of the commercial aspect of the business).

Each division was further divided into three/four subdivisions. The divisional manager was made responsible for the performance of the division. As a control measure, the divisional manager was required to make a presentation of monthly performance at the headquarter. The chairman and managing director of the company reviewed the performance. Changes like these:

— Stressed the commercial aspect of the utility business.
— Developed interrelationships between technical and finance executives.
— Enabled the divisional managers to understand the financial implications of the various activities undertaken.
— Involved the field personnel in various managerial decisions.
— Established an authentic database for important business parameters like billing, collection, T&D losses, etc., through a well-developed management accounting system.
— Increased accountability at the division level.
— Improved the performance at the division level.
— Made the top management aware of the field problems.

However, with privatization most of these activities came to standstill. Since most of the devolved activities were undertaken with an active support of the consultants, once the consultants were withdrawn from the process, it was difficult for the field staff to carry on the work.

10.1.3 Privatization

Orissa is the first state to privatize the distribution business by divesting more than 50 per cent of the government stake. The following were salient features of the privatization process:

— The Distribution zones were corporatized, giving birth to WESCO, NESCO, SOUTHCO, and CESCO in November 1998 as 100 per cent subsidiaries of GRIDCO.
— As per the transfer scheme of the Government of Orissa, distribution of fixed assets and corresponding equity was transferred to the distribution companies. In doing so, a relatively clean balance sheet was established for each of these companies, albeit, it had some negative effect on GRIDCO.
— More than 50 parties (both national and international organizations) had shown interest in the initial stages of the process. However, 12 companies/JVs submitted the required papers, of which 11 were pre-qualified by GRIDCO.
— To provide information about GRIDCO and the distribution companies, an Information Memorandum was given to the interested parties. GRIDCO also appointed separate committees for the evaluation of the technical and financial bids.
— Only three bids were received for three companies, viz., WESCO, NESCO, and SOUTHCO. For CESCO, a second round of bidding was invited.
— The process ended by transferring WESCO, NESCO, and SOUTHCO to BSES and CESCO to AES and Jyothi Structures Limited.

Even after two years of reforms there has been no significant improvement in the performance of distribution companies. Some of the observations are as follows:

— Though billing as percentage of the input increased marginally during the April-March 2000, still it was less than the OERC's approved 65 per cent. Overall billing was 56.8 per cent of the input in the year April-March 2000 against 54.0 per cent in the previous year. So the T&D loss remained in the range of 45-6 per cent, which is still considered to be very high.
— One of the important performance indicators of the distribution companies is 'Collection as % of Billing'. Overall collection has declined from 71.2 per cent (1998-9) to 64.4 per cent

TABLE 11.3: COLLECTION PERFORMANCE OF DISTRIBUTION COMPANIES

	Collection as per cent of Billing			
	WESCO	NESCO	SOUTHCO	CESCO
1998-9	90.9	86.3	78.4	75.9
1999-00	77.2	75.0	76.1	68.0

Source: Internal Records of the Distribution Companies.

(1999-00) for the LT and HT consumers. Table 11.3 shows such reduction in collection performance of the distribution companies

— The major shareholder of CESCO, AES is contemplating selling its stake and withdrawing from the distribution business.

11.2 DEVELOPING BUSINESS COMPETENCE USING VSM

For a problem of this magnitude, it is not possible to find solutions overnight. There is no quick fix to the problems facing the power distribution business or the SEB who are engaged in power distribution along with the other activities of transmission and generation. We need a paradigm shift from the conventional thought of changing the ownership for improving the performance, to developing organizational competence at the unit level. The focus should be on developing a viable business system. There are several models to develop such competence, some focus on the individuals and others on the organization.[10]

One such model is the Viable Systems Model (VSM). VSM of Stafford Beer[11] is based on ideas from the science of organization. VSM portrays a powerful image of a well-organized business and focuses on organization rather than structure (Jackson and Flood, 1991). A viable organization must be able to:[12]

— Make all normal decisions effectively;

— Adapt itself to changes in the demands made by the world (including customers, employees, suppliers, competitors, etc.) around it; and

— Learn from experience.

11.2.1 Principles of VSM

VSM is based on the following cybernetic principles (Jackson and Flood, 1991):

— The whole system is replicated in parts. In other words, an organization is a combination of several viable sub-systems and it operates within a larger system called meta-system.
— The sources of command and control are spread throughout the viable system, thus enhancing self-organization and localized management.
— The emphasis is on the relationship between the viable unit and its environment in terms of influencing and being influenced by it.

11.2.2 VSM: A Brief Introduction

The VSM is the arrangement of five functional elements called System 1 to System 5 that are interconnected through a complex information and control loop. The emphasis on repetition allows one to use the same basic model to represent, a company, its divisions and the wider organizations of which it may also be a functional part (Jackson and Flood, 1991).

The System 1 of an organization consists of the various parts of it directly concerned with implementation of the major objective/or the one responsible for undertaking the main activity of the organization. These are defined as departments, or divisions, or subsidiaries, or units etc. It is important to note that each part is autonomous in its own right. This means the parts themselves must be viable systems and must exhibit the features of all five functions.

Systems 2 to 5 are the channels for controlling (Holmberg, 1991) the separate System 1 and they emerge from the need to facilitate the operations of System 1, and to ensure the suitable adaptation of the whole organization.

System 2 comprises the information system necessary to decentralized decision-making within System 1 and to solving problems, which develop between the separate System 1s. This is carried out through formal reporting and through people building their own networks or contacts.

System 3: This is the control function that maintains internal stability and ensures effective implementation of policies.

System 4: An intelligence gathering function. It is the point in the organization where internal and external information can be brought together

System 5: It is responsible for policies. It balances today's operations against tomorrow's needs, and investments.

Besides the above five systems, the model stresses the importance of appropriate information flows and communication links. This information is about how the different parts of the organization, and the organization as a whole, are performing in relation to their respective goals. Beer (1985) suggests the following three levels of achievements:

— Actuality: Refers to the current achievement with existing resources within existing constraints;
— Capability: Refers to the possible achievement using existing resources with existing constraints; and
— Potentiality: Refers to what could be achieved by developing resources and removing constraints.

11.2.3 Application of VSM in Power Sector

Power sector is experiencing unprecedented changes because of the fast changes in the field of technology, feasibility of unbundling, changes in ownership and control, and dynamic goals of the different activities. In such a situation, the need for a scientific model, such as the VSM, that is based on cybernetic principles encompassing many ideas from management science cannot be overemphasized. Several authors have proved the general applicability of the model and it has been argued that the VSM can provide the premises to underpin efficient and effective managerial actions. VSM can help in diagnosing the problems facing the power sector and provide a mechanism to develop business competency.

In this section an effort will be made to use VSM as tool for developing business competence of one of the distribution companies, viz., CESCO by undertaking the following steps:

— Identifying the System;
— Establishing and studying five systems of the model at the corporate and division level; and
— Using VSM for addressing some of the problems facing CESCO.

11.2.4 Identify the System

The first requirement for applying VSM is the proper identification of the main system. This can be done by understanding the business activity undertaken by the organization. At present, CESCO is a licensee engaged in the purchase and sale of power in the central districts of the state of Orissa. The business operation is divided into around 15 accounting units. Each unit is responsible for undertaking the business and maintaining the necessary books of account. Each division is headed by an executive engineer, who is also in charge of the day-to-day operations. The divisional managers often look to the corporate office for directions on various managerial and operational decisions. The present organization structure of CESCO is shown by Figure 11.1.

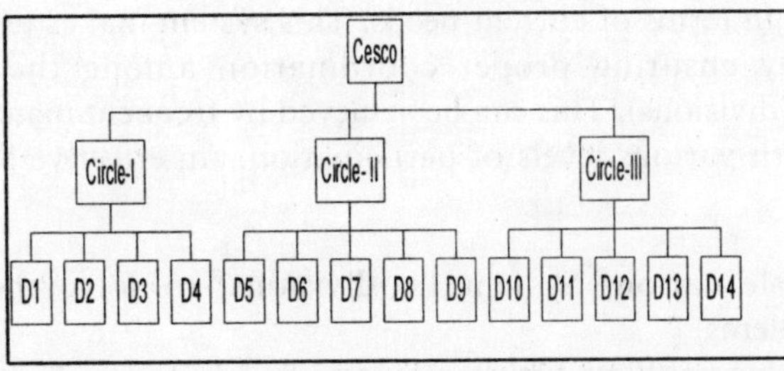

FIGURE 11.1: ORGANIZATIONAL STRUCTURE OF CESCO

Note: CESCO is divided into:

— 3 circles
— 14 divisions
— Each division is divided into 3-4 subdivisions

Some of the limitations of the conventional structure are as follows:

— The present structure is not stressing the main activities. The focus is on the flow of authority, i.e. top-down. Under this conventional structure, the budgets simply being imposed by central authority.
— Failure to delegate authority for resource deployment (decentralization in form but centralization spirit).
— Incongruency in the corporate and divisional goals.

The problem of centralization vs. decentralization is dealt with in VSM by allowing the sub-systems as much autonomy as is consistent with systemic cohesiveness. The metasystem , i.e. the top management will intervene only to prevent one of the parts acting in such a way that it threatens what the whole system is trying to achieve (Jackson and Flood, 1991).

Under VSM, a division will be treated as the System 1. So CESCO will be described as a combination of 15 System 1s. If CESCO has to be viable, it is necessary to ensure that all System 1s are viable. The main purpose of System 1 (division) can be defined as follows:

— To ensure power is supplied to the consumers
— To ensure all consumers are customers
— To utilize all resources optimally
— To ensure financial viability.

It is important to note that these systems are affected by the local and corporate environments. In order to maintain stability in the organization and address the environmental issues CESCO has to develop System 2.

System 2 refers to individual rules and behaviour for coordinating operations in terms of current needs. This system makes teamwork possible by ensuring proper coordination among the several System 1s (divisions). This can be achieved by frequent management reviews with various levels of participation. An effective System 2 will:

— Enable various System 1s (divisions) to solve their own problems
— Resolve conflicts between System 1s (divisions)

The channels of orders within CESCO can be called System 3. It can collect any information from any of the System 1s and feed the top management, or collect information from the top management and feed the System 1s. It helps in interpreting the policy decisions of the higher management and also carries out audits. The strong internal audit function of the organization can be treated as System 3. Most of the distribution companies (including CESCO) meticulously undertake the statutory audits and but fail to do justice to the internal audit. In the process, the very purpose of the internal audit is lost. So it is necessary to bring the audit function under

System 3. An effective System 2 and 3 will ensure internal stability and take care of most of the operating and day-to-day problems.

However, to keep CESCO viable and effective in the future, it is necessary to have a strong System 4. It develops capability to determine potential risks and benefits. A separate team chosen from various functional areas or a separate department designated as 'Brain' of company, can be asked to look after the activities of System 4. This group or the department identifies needs and potential for changes and informs System 3. The selected course of action can then be implemented by Systems 1 and 2 under the supervision of System 3. For example, the System 3 of CESCO can address the following managerial issues in the near future:

- — What will be the impact of Availability Tariff on its bottom line?
- — What will be the impact of multi-year tariff?
- — Should it buy power from GRIDCO or from the generators directly?
- — Is it necessary to undertake financial restructuring plan?

To understand the above changes and make the people capable of handling them, System 4 can initiate training programmes and create awareness.

System 5 completes the whole model. It is the advisory board of CESCO. This body balances the current activities against the future needs. Unlike the present system, wherein the board takes every decision, System 5 acts as the judge whenever there is conflict and makes policies on the basis of the feedback from Systems 3 and 4.

It is important to note that VSM lays stress on the autonomy of System 1s (i.e. on divisions and subdivisions). They are required to be viable in their own right taking into consideration their local environment. Systems 2 and 3 intervene to ensure overall cohesiveness. This allows the higher management to concentrate on the strategic and larger issues affecting the company and the sector as a whole.

To enable CESCO operate as a viable system, it is necessary to recognize the importance of information systems. VSM recognizes that information is the true cement holding the organization together. As per the requirement of the OERA, CESCO is required to file Annual Revenue Requirement (ARR), Tariff Application, and Annual Audited Accounts with the commission. The Systems 2 and 3 can handle this particular job efficiently. For this purpose, System 3 in association with System-2 can develop the following information systems:

— Capital Expenditure Information System (CEIS)
— Cash Management Information System (CMIS)
— Statutory Financial Information System (SFIS)

The top management can also use this information base for reviewing the performance of the divisions and use the same for taking various management decisions.

Recently the commission had asked all licensees to submit a five-year business plan. Such forecasting of the business and developing acceptable business scenario requires an in-depth understanding of the business and the changes that are affecting the sector. Such capability can be developed through System 4.

The fact that at present the annual accounts of CESCO, for the year ending 1999, are not ready and the lack of a robust business plan shows that Systems 2-5 are not functioning effectively.

At this stage, attention needs to be shifted to the field units, i.e. divisions and subdivisions. Are divisions operating efficiently? The answer is definitely not 'yes'. Only 40-5 per cent of the divisions are earning positive gross margin, most of the divisions are billing less than 50 per cent of the input, and very few divisions are collecting more than 70 per cent of the billing. Some of the causes for such a precarious performance are as follows:

— No effective information available at a local or company level;
— No uniform customer information and billing system;
— No reliance could be placed on meter readings;
— No credit control mechanisms;
— No mechanism to fix accountability;
— No effective performance measures; and
— No mechanism to involve the local stakeholders.

Though these problems have been identified in the early stages of reforms, even after the change in ownership they remain unsolved. So it is necessary to make these divisions viable to make CESCO viable. VSM provides the mechanism for addressing these issues by stressing the concept of strong and effective sub-systems.

Since each division generally consists of three to four subdivisions which can be treated as System 1s, it is necessary to have Systems 2 to 5 to coordinate with the subdivisions and also be in touch with the overall organization. Though the present structure has subdivision, it does not represent the features of System 1. They are created to facilitate administration rather than developing them as a viable system. Figure 11.2 shows division as a viable system.

It is important to note that in the present organizational structure, it may be possible to see the features of Systems 2-5 at the corporate level, however, they are conspicuous by their absence at the division level. As a result of their absence, divisions become dependent on the corporate office. Decisions taken by the corporate office are generally not in tune with the requirements of the divisions or the subdivisions.

One such casualty is budgetary control. It has been seen that the corporate objectives do not tally with the divisional objectives as

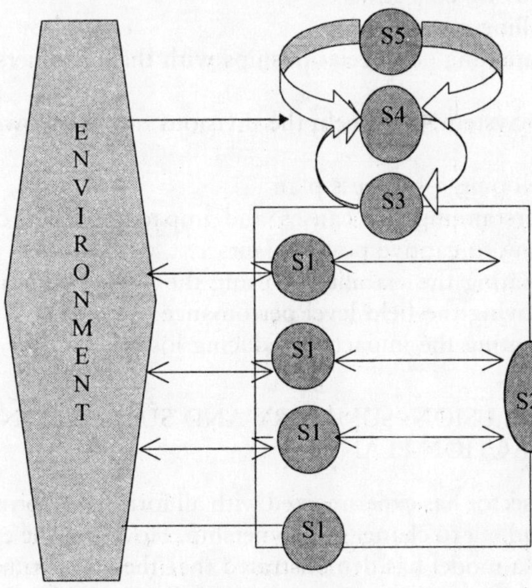

FIGURE 11.2: DIVISION AS VIABLE SYSTEM

- System 1: Subdivisions
- System 2: Rules and procedures to run the day-to-day operations at the sub-division level
- System 3: Coordination cell at the division office
- System 4: The brain or intelligence group of the division.
- System 5: Divisional management

also the targets given by the corporate office may not on the basis of the reality of the division or the subdivision. It has been pointed out that budgetary structure denies any actual delegation of discretionary power to the field units and the annual budgtets are still dependent on the discretion of the top management.[15] However, by shifting the management's focus to the divisions and subdivisions, it is possible to bridge this gap. Under VSM, the subdivisions or System 1 of the divisions will be responsible for making the budget on the basis of the local environment.

An effective System 2 and System 3 can help in undertaking the following activities more effectively:

— Establishing realistic budgets
— Reducing non-technical losses
— Improving collection
— Installing meters
— Maintaining good relationships with the customers

Similarly System 4 can help the divisions in the following ways:

— Developing a business plan
— Understanding the causes and impact of large customers shifting to captive power plants
— Evaluating the viability of using the village committees[16] in improving the field level perfomance
— Evaluating the impact of reducing losses.

11.3 CONCLUSION: SUMMARY AND SUGGESTIONS FOR ACTION PLAN

The power sector has experimented with all forms of reforms ranging from unbundling to changes in ownership. However, the experience of the Orissa model has demonstrated that these experiments have not changed the scenario. The sector is still affected by most of the problems and is now facing newer problems arising from the failure of the privatization process. This paper presented an alternative model for strengthening the organization by developing the business competence using VSM. VSM offers a paradigm for problem-solving. While no attempt has been made to find solutions to specific problems, an effort has been made to diagnose the limitations of the existing solutions. It has been suggested that it necessary to see the organization, i.e. CESCO as a Viable System consisting of several viable systems. It has been observed that though some form of all five systems can be found at the corporate level, such systems are conspicuous by absence at the division level. If each subdivision and division becomes viable then the entire organization can become viable too.

NOTES

1. The Power and Energy Division of the Planning Commission, New Delhi in their Annual Report on the Working of SEBs and Electricity Department (April 2000), reported the T&D losses as 23.5 per cent (1992-3), 23.4 per cent (1993-4), 23.8 per cent (1994-5), 46.9 per cent (1995-6) and 20.4 per cent (96-7).
2. S. Hunt and G. Shuttleworth, *Competition and Choice in Electricity*, Chichester: John Wiley & Sons, 1996.
3. The generating plants were large with respect to the size of the market. The optimal size of the generating unit rose through 1970 to 1980, as it had for the previous 50 years. Then things began to turn around. Technology imported from material science and space programmes made turbines much more efficient. At the same time the price of gas declined. The path was then open for smaller and cheaper generating units.
4. J. Surrey, ed., *The British Electricity Experiment, Privatisation: The Record, The Issues, The Lessons*, London: Earthscan Publications, 1996
5. D. Farnham, and S. Horton, 'The Public Service Managerialism', in D. Farnham and S. Horton, eds., *Managing the New Public Services*, London: Macmillan, 1993.
6. N. Flynn, *Public Sector Management*, Harvester: Hemel Hempstead, 1993.
7. G. Stoker, *New Management Trends*, Luton: Local Government Training Board, 1989.
8. L.D. Terry, 'Administrative Leadership, Neo Managerialism, and the Public Management Movement', *Public Administration Review*, vol. 58, no. 3, 1998, pp. 194-200.
9. D. Osbrone and T. Gaebler, *Reinventing Government*, New York: Addison Wesley, 1992.
10. For details on various models reference can be made to *Creative Problem Solving: Total Systems Intervention* by M.C. Jackson and R.L. Flood, Chichester: John Wiley & Sons, 1991.
11. Stafford Beer unfolded his model (VSM) of the organizational structure of viable systems in two of his books, viz., *Brain of the Firm* and *The Heart of Enterprise* published during the 1970s. He published *Diagnosing the System for Organisations* in 1985, which was intended as a practical guide to enable any manger to organize and manage tasks effectively.

12. B.A. Holmberg, 'Developing Organisational Competence in a business', in R. Espejo and R. Harnden, eds., *The Viable System Model: Interpretation and Applications of Stafford Beer's VSM*, Chichester: John Wiley & Sons, 1990.
13. R. Espejo (1979), 'Information and Management: The cybernetics of a small company', working paper 125, University of Aston Management Centre; M.C. Jackson and B.O. Alabi (1986), 'Viable systems all: a diagnosis for XY Entertainments', working paper 9, Dept. of Management Systems and Sciences, University of Hull; S. Beer (1974), 'The Integration of Government Planning', Study for the Government of Alberta.
14. M.C. Jackson, 'Evaluating the Managerial Significance of the VSM', in R. Espejo and R. Harnden, eds., *The Viable System Model: Interpretation and Applications of Stafford Beer's VSM*, Chichester: John Wiley & Sons, 1990.
15. Joël Ruet, 'Delegation of Power in SEBs under Management Contracts', paper presented in seminar on 'Organizational Restructuring of SEBs', New Delhi, 2001.
16. The concept of village committees has been used in Orissa for improving the billing and collection performance in some of the divisions of WESCO and NESCO.

12

Practical Proposals for SEBs

JOËL RUET

The concluding chapter presents the proposals that have emerged from among the various contributions to this book, and details some of them. The contributors, have collectively evolved a set of proposals, which are worth being tested in the field. And though each individual contributor does do not necessarily fully subscribe to all of the proposals, the essays taken as a whole do display a certain consistency, enough anyway to give us a chance of overcoming the present impasse.

12.1 THE SET OF MEASURES

The proposals are broken-up along (i) operational, (ii) managerial, and (iii) regulatory measures, and provide a self-consistent framework for reform. Besides, most of them are viable in themselves. Together, they provide positive externalities and virtuous circles, and are definitely profitable.

12.1.1 Operation/Technical

(a) Institutionalize metering for all consumers, and convert billing on average into billing on actuals.

On selected areas, a socio-economic study (energy uses and willingness to pay), combined with an organizational study, will provide the outcomes of such a measure in terms of (i) increase of the maintenance budget for the utility, and thus (ii) positive effects of higher power availability in respect to the high cost of coping strategies (investment in captive generation).

(b) Outsource auxiliary services

Existing studies on the positive impact that increasing the availability of reactive power has on the investment and operation costs should be (i) extended to some pilot zones and (ii) encouraged as a macro-economic approach.

(c) Develop benchmarking techniques for evaluation of SEBs' efficiency.

Both gathering the 'best practices' in SEBs and developing analytical and technical tools for the evaluation of their efficiency would enhance technical operation tools available within SEBs, as well as enable a benchmarking useful for the regulation of the sector.

12.1.2 Management

(a) Develop and clarify implementation of internal audit. The concept should shift from transaction to system audit, and also from statutory financial audit to management audit.

Pilot studies should focus on identifying and phasing out the structural changes in the internal information system: avoiding statistical entries and generalizing accountancy entries; integrating the reporting (bringing together physical measures, accounting and finance); identifying the right level for such a system audit to be implemented, and identifying the areas of auditing, given (i) the present internal allocation of powers, (ii) considering a decentralization of powers as proposed in the current Electricity Bill.

(b) Establish clear internal guidelines on duties and responsibilities, and develop the commercial function at the divisional level.

Pilot studies should focus on the necessary upgradations/ simplifications and implementation of transitory ad hoc monitoring committees in the reporting system.

(c) Develop viable business units based on the concepts of decentralization and profit-centres.

It is proposed to provide an identification, analysis, and segregation of the types of costs to be considered in establishing a profit-centre scheme. The optimal phasing out of such an implementation would also have to be studied. The need for a suitable delegation of power shall be closely examined.

(d) Strengthen and actually implement a delegation of power in the use of resources.

This measure can be complementary to the precedent. But it may as well be implemented in itself, and a pilot study is needed to identify the margins of internal resource generation that can come from a better operation allowed by decentralization (preventive maintenance on distribution transformers for instance).

(e) If privatization is an option, the best way of going about it is through management contracts, with specific legal and financial provisions on hiring external people and commitment of financial resources to optimize revenues.

An in-depth study of the privatization in Orissa and of other privatizations currently in progress will help in the development of action-oriented models. These should focus on the pros and cons of direct and quick privatization, and provide methodological tools regarding optimal conditions and right timing to engage in privatization, if such is the choice of a state government.

12.1.3 Regulation and Institutional Environment

(a) Reform does not necessarily mean either beginning or ending with unbundling of SEBs.

Field study of unbundled SEBs will indicate the steps that remain to be taken for a full and actual unbundling, and further show the need for other decentralization measures than unbundling per se.

(b) Promote open access and wholesale competition on multi-buyer model.

An extensive study should be carried out on the institutional and organizational steps that need to be realized in order to achieve a multi-buyer model.

(c) Remove current distortion between NTPC and IPPs in financial returns and guarantees.

A macro-economic study is proposed on modelling the respective effects on redistribution and investment of various scenarios of regulation convergence between NTPC and IPPs.

(d) Review guarantees for IPPs, create an environment where IPPs will be glad to take the business risks.

Taking stock of the present absence of sustained interest from the IPPs, an independent report on their perception of the business environment is proposed.

(e) Encourage regulatory commissions to go for price-based regulation rather than cost-based regulation (or the RPI minus X variety), and explore other market-based pricing mechanisms; and encourage regulatory commissions to give multi-year formula and rules in tariff fixing.

A survey of the implementation of this mechanism in various countries (including England and Wales), and of other systems (including the French system) will enhance the public debate on regulatory commissions.

(f) Assess the feasibility and merits of direct subsidies.

Pros and cons of direct subsidies and targeted subsidies shall be assessed through both a field study, as well as a consumer-oriented and organizational analysis indicating the steps to be achieved in the restructuring of the SEBs in order to implement such a scheme.

12.1.4 Rationale in the Set of Measures

These measures are designed to offer a specific advantage if seen as a unity. The technical measures aim at ensuring quick visible results for the consumer and the SEBs' agents, to generate a supportive atmosphere for the reforms, and allow both managerial and regulatory steps. The managerial proposal will give powers and means to field agents for implementing these technical and commercial measures in the most targeted manner. Also, a phased delegation of power will ultimately free the boards from the day to day problems and allow them to concentrate on restructuring issues, the evolution of the managerial organization, as well as evaluation of technical schemes. On the top of the structure, sound dialogue on loss reduction, as well as an actual competition in a fair environment are promoted. Under a clear and lasting framework, this allows for benchmarking and transparent evaluation of delegation schemes, along with the different incentives, for both the board and the field officers.

12.2 PRACTICAL IMPLEMENTATION OF SOME MEASURES

This section sums up the practical proposals made in various chapters of the book.

12.2.1 Development of Internal Audit

The audit committee should be generalized in a proper way: such a committee normally comprises the members of senior management of an enterprise who oversee and review the operations of IA functioning. Such committees are necessary in the case of huge enterprises in terms of operations and investment, requiring these to have an effective view of operations and to undertake an examination to ascertain the effectiveness of the control systems. These committees are normally set up by the board and may comprise members drawn from the disciplines of Accounting, Economics and Law. The committee might also have experts with an intimate knowledge of the functioning of the industry of which the firm is a constituent. The recognition of the need to bring the IA function under the control of audit committee implies the great deal of importance that an enterprise attaches to the audit function. Also, the presence of a strategic audit plan is a primary requirement for the success of the IA function in an organization. Such a plan should spell out the duration, audit priorities, and the target maximum audit cycle.

At a time when privatization of the activities of the SEBs is gaining ground in the country, it might be worthwhile to consider whether the IA function could be contracted out to produce better results. SEBs should be granted rights to recruit and appoint well-qualified and experienced personnel. A continuous training of the IA staff is also suggested.

12.2.2 Establishing Profit Centres

Requiring the board to have meters for all feeders, both at incoming and outgoing levels for each division, could eliminate manipulations on loss figures, as well as identify the culprits who are responsible for the revenue leakage. The board could view each division as a profit centre (reporting details on energy: received, metered, sold; on revenues: billable, billed, realized, deficit in billing, current arrears, loss, cumulated arrears). By circulating the division-wise performance to all divisions, every division chief is assessed by all other division chiefs and this creates what in audit parlance is called audit anticipation effect.

The second step is to arrive at a concept of surplus for each division. Surplus is nothing but cash inflow minus (cost of received electricity plus out of pocket cash outflow for the division) for the month. The cash inflows are proceeds through sale of electricity. The cost of received electricity is the product of the number of units of electricity received by the division and the transfer price of electricity received at the pooled cost of generation and purchase, and an adjustment factor for the load mix. One can call this the load mix factor. The load mix factor will compensate for the lower tariff loads by reducing the input cost of power for the division. To arrive at the surplus, division-wise energy accounting is a necessary condition. Now this surplus figure too could be circulated to all divisions. Divisions that make a surplus beyond a stipulated amount could be given resource discretion. That is, they could be permitted to retain the surplus for their budgeted and authorized expenditure, instead of repatriating the entire revenue to the head office and then waiting for the expenditure amount to come from the head office.

The third and final step is to establish each division as a profit-centre clearly providing them the depreciation figures of their assets and the overhead allotted. Once the profit for each division is calculable, then a financial incentive system could be developed so that every one in the division is rewarded for their efforts towards creating the profit, in proportion to their efforts. Having established such a system, three measures should be taken care of:

— Greater incentives for field people would be a step in the right direction. After that, however, the total SEB's performance based incentives could be extended to the head-office staff. Incentives for common services could be worked out in their transfer price. At the same time, the operating divisions could be allowed freedom in deploying their services based on their own cost benefit of the services offered, on a competitive basis.

— Non-performing divisions could be handled by giving them a long rope, by management resorting to judicious transfers, and isolating the effects of bad employees by making them redundant, etc.
— As regards the guarantee that the incentive system would work if the existing system promised more rewards, the entire incentive-based profit-centred approach is designed upon the premise that marginal utility of earned incentives would be more than marginal utility of ill-gotten income for most employees.

12.2.3 Benchmarking the SEBs' Activities and Outsourcing the Auxiliary Services

Local performance indices (LPI) should be developed to benchmark each activity in the organization. For example, LPIs for generation may include the residual life of a generating unit, environmental impact, nature of combustion technology, and maintenance costs. Similarly, LPIs for distribution may consist of losses, reactive power requirement, connected load, projected connected load, life of major equipment such as transformers, power factor and/or displacement factor. Line availability, overload capacity, compensation mechanisms, and voltage levels as well as losses would constitute the LPIs of the transmission system. Finally, the LPIs should be combined to arrive at a global performance index (GPI), akin to a credit rating, for each SEB.

SEBs can be unbundled into four services: Generation, Transmission, Distribution and Auxiliary Services. The natural choice when initiating privatization is to start with privatization of auxiliary services, primarily for the following reasons:

— Total investment made by the private investor would have been minimum.
— Improvement of voltage profile and increased power availability can build customer confidence for privatization. Farmers have often demanded greater availability of power, even at a cost, instead of irregular and poor quality free power.
— Transition of SEBs to the market environment would be gradual. Moreover, introduction of auxiliary services would in any case improve the system strength.

Once system strength is improved the privatization would be enhanced. Benchmarking of auxiliary services is not a difficult task. Auxiliary services can be treated in a manner similar to generation. The only difference is that in generation the primary objective is to meet real power demand whereas auxiliary services would provide support to reactive power and filter for harmonics. These services can easily enhance system capability up to 20-30 per cent. Such an approach would have sent signals even to the common man on street that we are thinking of improving his life rather that snatching away his free lunch.

12.2.4 Delegation of Financial Power and Management Contract

In SEBs, lack of budgetary power precludes coping with technical stakes like efficient maintenance of the network for instance. Two methods for delegation of financial power are possible. One way is through 'direct allocation'; headquarters should start giving cash allocation to the field, regardless of initial improvements in revenue collection. The other way is through 'bound allocation', by waiting for initial revenue recoveries.

The cash allocation has to be done according to a given timetable. To start the reform, it could concern investments, maintenance, and repair of distribution transformers. Let the headquarters authorize the field to use:

— either this cash in these transformer-related fields, irrespective of the share between them, i.e. to divert from previous use within these budget lines.
— or to use a limited part (up to a fixed limit, say a few per cent) of this cash for any input purchasing.

This will enable the field to use the cash where it has information on its efficiency, lowering the working charge of headquarters and improving reactivity. For the headquarters to be sure to control the total budget, it may fix the maximum limit on it. As far as revenue is concerned, let the headquarters specify a minimal revenue collection, based on feasible targets, for instance previous achievements. To be credible, the headquarters can make their cash disbursement subject to completion of a given timetable in collection. Over and above this, let the headquarters allow the field to retain, as an incentive, a part of the extra realization for its own cash account (this measure can be additional to a personal incentive scheme).

PRACTICAL ISSUES TO BE SOLVED

- Setting up of this system needs to happen in the following framework:
  - (i) A transparent and synthesized information system is in place, allowing both managerial tools for the field and control tools for the headquarters.
  - (ii) A coordinating committee between the field and headquarters is established, to allow the monitoring of budget constraints.
  - (iii) Training of the SE and the Xens is organized, with accounting support from the headquarters.
- In particular, corruption problems might occur. Under this system, the money would not be easy to divert directly to personal accounts, as there will be some control. However, a mechanism needs to be in place, to ensure that managers do not enjoy it physically, by buying unnecessary inputs. This percentage allowed for inputs purchasing might either be introduced after a while, or could be under control of the monitoring committee through specific reports on their final profitability.
- A priori, direct allocation shows a higher commitment of the headquarters to the field, but the bound one offers an incentive to make collections faster. A mix of both, or initiating the reform with direct and then replacing it with bound allocation, could be the best way out.
- The introduction of cash at the circle level also implies the introduction of cash in the store and workshops, with guarantees given on timetable and costs, and/or authorization to purchase from the private sector in case of non-compliance with the commitment (under the ex-post control of the coordinating committee).

In that process, the public sector can benefit from private experience. Management contracts can be drawn up, with the following proposals to ensure credibility:

— make provisions for the private sector to administratively sanction the faults, at least by posting different people in place of corrupt officers,
— give the right and means to add privately-financed posts in the hierarchy, with the right to invite the public officers to join this privately-financed type of posts (as tried by CESCO in Orissa),
— grant full discretion over a certain amount of investment,
— base the fee both on a fixed part and on a success-based part linked with the operating surplus to sales in current prices.

12.2.5 Financial Links with the State Government, and Agreement with the Regulatory Commission

The Orissa case shows that, DISTCOMs have taken several initiatives to improve collection and reduce losses, that can be considered as 'ideal practices'. The most important of these has been changing the work culture to introduce a commercial orientation among the employees. For instance, CESCO has taken some bold steps including open-market recruitment for professionals (about 50) into their 'business cadre', assigning accountability and powers to managers and workers and taking strong disciplinary measures for non-compliance. Other steps to improve efficiency include:

— Disconnections to force consumers to pay their bills.
— Bijali Adalats and consumer camps to provide speedy redressal of complaints.
— Establishment of vigilance wing to detect theft and malpractice, and administrative action against erring employees.
— Establishment of customer care centres.
— Village committees

The village committee is a committee of consumers from the village with participation of field staff of the utility with the objective of facilitating a single-point contact between the utility and the consumers of the village. After a preliminary phase, the scheme was extended to an additional 3,000 villages in the NESCO and WESCO areas. In 1,000 of these villages the second phase of the initiative has been launched whereby greater accountability is proposed to be transferred to the committee by making them aware of the input at transformer level and actual billing being achieved from the village. All the consumers in the village are proposed to be metered in this phase. In the third phase it is expected that energy accounting will be entirely transferred to the village committee whereby they will perform like a cooperative, i.e. they will be billed based on transformer readings and entire responsibility for collections will be transferred to them.

However, after the three crucial stages of the reform process—corporatization, unbundling and privatization of distribution, and after the setting up of the regulatory body, the sector is however faced with a serious financial crisis. This is rooted in the mismanagement of T&D losses, and in the relationship between regulated tariffs and T&D losses. The only way out of the current stalemate—and if similar ones are to be avoided in other states— is for the regulatory commission, the licensees and the state government to jointly agree to:

— a multi-year sustainable tariff schedule,
— set of efficiency improvement targets,
— and phased out annual subsidies from the government to the licensees which ensures that the licensees break-even.

Especially, multi-year tariffs are essential to enable the various players in the power sector to plan for the future with some degree of certainty. Multi-year tariffs will also involve pass-through of power purchase costs. Apart from fuel costs, these other costs which are beyond the control of the licensees will also have to be passed through.

The tariffs should also provide for a phased reduction and ultimate elimination of cross subsidies. In case the government wants to provide subsidies to certain segments these should be clearly targeted and paid for separately by the government.

Licensees will have to provide a multi-year schedule of progress towards achieving the standards on which the tariffs are based. For this schedule to be meaningful the efficiency parameters should be measurable. This requires a level of metering, not necessarily 100 per cent, which will provide reasonably reliable T&D loss data.

Given the multi-year tariff schedule and the multi-year programme of efficiency improvements the cash deficits of the licensees can be worked out. The cash deficits represent cost of the gap between standard costs and efficiency parameters and the actual costs and efficiency parameters. This cash deficit will have to be subsidized by the government. Other alternatives such as creation of regulatory assets are unlikely to be credible. The government could levy a special temporary tax to finance this subsidy. Support to this measure comes from the fact that, although not formulated in this manner, the approach in Andhra has some elements of this proposal. The Maharashtra regulator also appears to be moving towards this arrangement.

12.2.6 Incentives on Cost Saving, Tariffs and Subsidies

Capping the subsidy, and separation of the administration of subsidy from the power entities are important issues. Thus, stopping the leakages and improving efficiency can no longer be ignored.

Reform should be defined as a reorganization of sorts. It is imperative to bring about legal and policy changes which would:

- — solve the agency and moral hazard problem in the administration of subsidies;
- — create incentives for cost saving and efficient generation, transmission and distribution;
- — correct the inversions in the tariff, and lead to a scheme that would allow prices to play their economic role of reflecting costs, and being a signal for outputs to respond.

As far as Orissa is concerned, the proposals can be summed up as follows:

TRANSCO should lay down detailed guidelines, announce connection, congestion and line-loss charges for various types and sizes of loads connected to the transmission network and for use of its facilities, in consultation with the regulator. The assets would have to be valued at replacement cost minus the maintenance costs required to bring them to a state of complete repair.

After unbundling and disinvestment, with less than majority shares held by government, the company would not be 'state' in the courts' view and would have the requisite freedom to pay its staff well enough and in a manner that creates incentives within the organization for performance.

Privatizing the existing distribution zones would not pose any problem when the subsidization is direct. There will be variations in distribution costs, but, if the regulator adopts ROI type regulation, it would mean somewhat different ceiling tariffs to the final consumer. The differences would reflect the varying average cost of distribution alone, arising out of the variation in consumer mix across zones.

To arrive at the base cost for distribution, its indexation, or the return, a detailed study of distribution cost (variable and fixed) as a function of the various measures of physical network like the density and type of consumers, would be necessary. With ROI type regulation, an announcement that the distribution and generation companies would on their replacement value of assets, earn a return of 16 per cent, would remove any regulatory risk in the sale of both generation and distribution assets. It would then be the regulator's annual task to approve the tariff demands placed upon it by the DCGs. With price-cap regulation, the regulator would have to announce his formulae of prices for various cost segments.

The best choice for the regulator would of course be a price-cap approach to setting prices for different components of transmission and distribution costs: fixed, variable and penalties in exceeding designated loads.

Allowing all loads at 1MW or above and all CPs with installed capacity 1MW or above, the right of access to the transmission/ distribution lines would be the key to bringing about competitive pressure. For such customers/entities, the sale and purchase of power is completely separated from its carriage.

The rules for market making would have to be spelt out by the TRANSCO, after system and market simulation studies. These would also alert it to possible pitfalls. But in the meantime simple rules that insist upon matched contracts for sale and supply with less than 5 per cent variation, before allowing use of the grid by all participants, would serve the purpose. They would of course have to be checked for systemic feasibility—given the grid capacities—through a simple programme.

12.3 CONCLUSION: AN ANALYTICAL SHIFT

This book, and the diversity of its proposals—all measures to be tested as solutions to the current impasse in power sector reform, shows the present analytical shift at stake in India. Indeed, a growing number of voices, including those expressed here, claim to address the issue of reform in new ways and with new priorities. They criticize the dominant and schematic public-private debate as a debate per se, and emphasize the need for an actual change in the internal organization of SEBs. They advocate going beyond the classical tools of corporatization and deintegration without substantial practical effect. The proposed solutions encompass a gradation from cost centres to profit centres, and finally decentralization. They all propose measures for autonomization of the operational units; therein lies the crux of all decisions affecting efficiency. Articulated to these internal measures, the issue of privatization is renewed through the specific and targeted involvement of the private sector.

Be it in auxiliary services, or in operation through management contracts, or the building of actual trading markets, the private sector is more and more regarded as a specific provider of different inputs. This can range from investments in some cases, to managerial experience in others, or even to a catalytic impetus for further reforms. Briefly, the analytical tools provided in this book are diverse and go far beyond clear-cut simplistic approaches, since reality is diverse and complex.

These approaches offer more than mere intellectual appeal; practical implementing processes are proposed as well, since practical experience has accumulated enough for this. Compared to present reforms, they offer the advantage of modifying the day-to-day functioning of the organization. In particular, they try and look beyond mere cosmetic surgery for unchanged practices. In contrast to the privatization credo—where one has had to acknowledge that investors do not come rushing in—they highlight the present administrative functioning of SEBs, and the structural changes that need to be brought about therein.

Last but not the least: all these proposals are open to debate and should ultimately be subjected to the test of the field. But the nature of the analysis is definitely changing.